After an estimated 10,000 attempts, Thomas Alva Edison invents the lightbulb in 1879. At the same time, Englishman Benjamin Swan invents it too.

THE ILLUSTRATED TIMELINE OF
Inventions

THE ILLUSTRATED TIMELINE OF Inventions

A CRASH COURSE IN WORDS & PICTURES

Craig Sandler

New York / London
www.sterlingpublishing.com
A JOHN BOSWELL ASSOCIATES BOOK

Library of Congress Cataloging-in-Publication Data Available

2 4 6 8 10 9 7 5 3 1

Published by Sterling Publishing Co., Inc.
387 Park Avenue South, New York, NY 10016

Distributed in Canada by Sterling Publishing
c/o Canadian Manda Group, 165 Dufferin Street
Toronto, Ontario, Canada M6K 3H6
Distributed in the United Kingdom by GMC Distribution Services
Castle Place, 166 High Street, Lewes, East Sussex, England BN7 1XU
Distributed in Australia by Capricorn Link (Australia) Pty. Ltd.
P.O. Box 704, Windsor, NSW 2756, Australia

Printed in China.

Sterling ISBN-13: 978-1-4027-4862-2
ISBN-10: 1-4027-4862-0

For information about custom editions, special sales, premium and
corporate purchases, please contact Sterling Special Sales
Department at 800-805-5489 or specialsales@sterlingpub.com.

Book Design by Barbara Aronica-Buck

Contents

Acknowledgments

The wonderful idea for this book wasn't mine, so thanks are absolutely due first to John Boswell, Christa Bourg, and the others at John Boswell Associates who brought the series into being and moved it along so expertly. Designer Barbara Aronica-Buck did an amazing job capturing both the scope and detail of technology's advance. Melanie Gold's good humor, thoughtful and sensitive feedback, and joy in the endeavor provided a standard for professionalism and kindness by which I'll judge all other editors from now on. Sterling turns out to be the perfect name for the company, given the qualities of the people there and their work.

In truth, I've been surrounded by role models during the production of this book, not all of whom I can mention, and I hope they will forgive the omission.

Martha, my wife, responded brilliantly as I challenged her with yet another opportunity to display the patience, warmth, and depth of character I cherish. Her father, Charlie, provided his usual clear thinking and good spirit as he lent important feedback on entries dealing with physics.

Above all, though, there's the example of my dad, who first brought this project to me and me to it but, much more than that, spent a lifetime imbuing me with his love of words and his passion and respect for the people of the past and the craft of telling their stories. His skill, wisdom, and love would have made him the ideal mentor for this project even if he hadn't been born on Inventor's Day. Nothing gives me more pleasure than the prospect of continuing to follow his joyful footsteps.

For Lila

Introduction

Why should you care about the history of inventions? Mainly, to appreciate how our world and our daily lives came to be. Practically every item you use, from the ice cube to the Internet, was first a possibility in someone's mind, then made reality through skill and persistence. Over a few million years almost everything that surrounds us has been created this way.

Every invention is the solution to a problem, and studying the vast and varied history of solutions yields a deep new appreciation of—and perhaps even faith in—humankind's ability to grapple with and overcome its challenges.

Finally, looking under the hood of the world we take for granted reveals a cavalcade of captivating stories, and nothing is more fun than learning them. Wilhelm Röntgen switches on a vacuum tube in his lab, and for no apparent reason a container across the room begins to glow; before you know it, doctors are saving lives with X-rays. William Hunt is challenged by a pal: if he invents something useful with one piece of wire, the friend will cancel a debt. Now we can thank Hunt for dreaming up the safety pin. The whole history of invention abounds with these amazing stories—all the more entertaining because they're true.

With the exception of language, no other trait so defines our human species—this fierce instinct to solve problems with our heads and then our hands. The process of transforming problems into new ideas and new ideas into solutions: we call it *invention*.

But it's the second part of that process that truly has an impact on the world. It's almost unheard of for the same person to come up with both an original idea and the practical working version of the idea. From building fires to building lasers, the fame and glory are reserved mostly for those who first imagine making something happen, but those who first figure out how to actually make it happen.

Bringing someone's idea to life is fiendishly difficult, its challenges surmountable only by the time-honored virtues cited by every great inventor: persistence, hard work, determination. No one before or since has expressed it more eloquently than the American genius

Thomas Edison. Having failed ten thousand times in his quest to devise a working lightbulb, he told a reporter he'd actually succeeded in identifying ten thousand ways that wouldn't work.

So, inventiveness in the true sense means not just intellect, but a kind of genius for perseverance, for practicality. "I learned pretty early in life that you don't have to learn everything to be able to do something. With inventing, you are attempting to solve a problem within your reach, not trying to resolve the world's greatest problems," said Nick Holonyak, inventor of the light-emitting diode and holder of thirty-one patents. He expressed the spirit of an odd and unpredictable realm of endeavor: part brilliance, part drudgery, and part luck; neither art nor science, but containing elements of both in a combination that's utterly transformed our planet and someday could destroy it.

Every day, new inventions appear, adding to the roll call of dreams turned into reality. And each one adds to our ability to innovate further, making it possible for even bigger, stranger, more exciting dreams to come true. This book is the record so far.

How to Use This Book

This timeline is divided into four sections, but if you look carefully you'll notice how they reflect the accelerating pace of change, discovery, and inventiveness. Though the first and last sections are about the same length in pages, the first covers about two million years and the last only a hundred. Each successive section covers a shorter amount of time, yet each contains roughly the same number of inventions.

Entries along the central timeline convey different types of information. Most frequently and obviously, the appearance of a specific device or process is chronicled. But other entries note the discovery or insight that enabled subsequent, world-changing inventions. You'll be able to detect threads and cycles, too; for instance, the microscope, an invention, led to germ theory, a scientific principle; and that led to inoculation, another invention.

Entry texts are color-coded to show you, at a glance, where in the world each discovery or advance occurred. Sidebars are scattered throughout the book to provide more detail—a bit more about the life story of an inventor, or the social setting or impact of a technology, or simply a fun fact. This accessible, visual presentation, designed to facilitate browsing, will help you understand and enjoy the amazing chronology of inventions, whether the book serves as an introduction or a refresher course.

Color Key

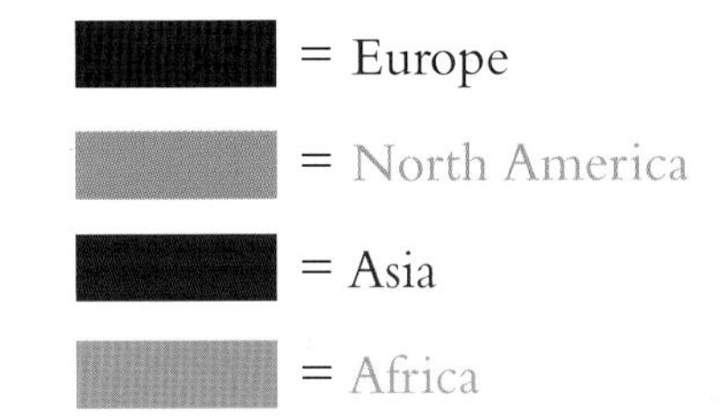

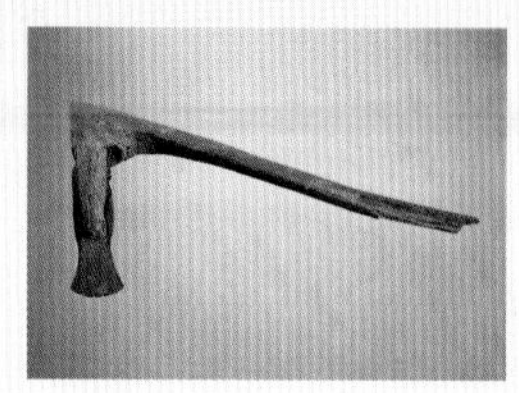

From the Spear to the Printing Press: 1 Million BC to 1447 AD

Necessity is said to be the mother of invention, and the greatest inventions arose from the greatest necessity, in the predawn of the human race. Millions of years ago, on the plains of Africa, certain hominoids thought of new and better ways to meet the brutish challenges of prehistoric living. These individuals stood out, survived better, and gained an evolutionary advantage. The ability to invent joined the ability to speak and reason in the center of the human essence.

Humans did not invent the tool, or the control of fire—our biological ancestors did. *Homo habilis* (2.5 million BC) means "handy man," for this species' ability to create hand axes. As its successors—*Homo erectus* (Neanderthal man) and *Homo sapiens* about one hundred thousand years ago—increased in brain size, each new species superseded the last by outthinking the competition. The last Neanderthal man (or woman) died about thirty thousand years ago.

By the time Neanderthals disappeared, humankind already had invented glue, lamps, needles, and musical instruments. The day was approaching when someone would invent a way to make fire, thereby freeing humankind from having to safeguard precious embers gleaned from natural happenstance.

In well-established homes in Southern Asia and Mesopotamia, *Homo sapiens* mastered the power of fire, which allowed them to disperse northward. Observation, experimentation, and insight, combined with courage, persistence, and good luck, slowly yielded one new improvement after another. As with all primeval innovations, placing a reliable date on that breakthrough may be permanently impossible; the dates shift as archaeologists dig deeper, so to speak.

What does seem clear (as of today) is that making fire cleared the way (literally and figuratively) for sowing seeds—learning to intentionally cultivate and harvest crops for food. The earliest traces of cultivation have been found in Africa and Asia and date from about 10,000 BC.

Having a reliable hearth around which to gather and a semireliable source of nourishment provided the opportunity to experiment and reflect. The emergence of pottery is simultaneous with agriculture in the archaeological record, and copper working—a result of the same urge to study how materials behave when heated—inevitably followed. Most important, the settling-in associated with agriculture permitted tribes to become villages, then cities, and then kingdoms. The greater sharing of information that comes with urban life increased the pace of advances in transportation, communication, adornment, and material well-being.

These developments were dispersed from Atlantic to Pacific, in an Old World sense: from Chardonnay to China, Denmark to Damascus. The harpoon is found first in France; agriculture started in Mesopotamia; the first pottery came from Japan. No area of the world had a franchise on turning out innovations before anyone else. But clearly, collective genius did concentrate at certain points on the ancient map: China, Egypt, and Mesopotamia, the home of successive civilizations (Sumer, Babylon, Assyria) between the Tigris and Euphrates rivers.

These great civilizations and others, such as Anatolia in modern-day Turkey, served as wellsprings for the next wave of fundamental technologies after agriculture and urban living were established. From Anatolia came remarkable new contrivances like the mirror and the map. From the great city-states of the Middle East came the inventions of cloth weaving and eventually bronze making. The Indus Valley civilization along the Ganges gave rise to inventions from ink to integers. Ancient Egypt from 5000 BC to the first millennium gave the world a staggering array of items it can no longer live without: the calendar, irrigation, the scale, cement, the sail, glass, the astrolabe, and the lighthouse. Meanwhile, the Mesopotamians brought forth the icon of invention, the wheel, along with a plethora of brilliant essentials: writing, the canal, the pulley, soap, wine, and beer.

These innovations were for the most part spread by traders—mariners, mainly—journeying to other lands in the hunt for new markets for goods produced and harvested at home. China, without well-plied trade routes and adhering to a conscious strategy of keeping proprietary technologies like silk making and gunpowder secret as long as possible, made astonishing technical progress on its own for millennia; moveable type four hundred years before Johannes Gutenberg is just one example. The more mobile civilizations and kingdoms tended to impart more of their knowledge and inventions to the rest of the world, and acquire more new ideas as well. This is part of the reason Greece developed into the greatest engine of ingenuity the world had yet seen. Rome with its far-flung journeys of conquest followed suit. After Rome fell, Europe awaited the next wave of journeymen, the Arabs, to arouse it from a long intellectual slumber; the ideas spread from Arab empires were the seeds of the Renaissance, which came into true first flower with the invention of the Gutenberg printing press.

Shortly after Gutenberg's work transformed civic life in Europe, Christopher Columbus encountered the Americas, where he and his successors found a culture that had a whole slate of excellent ideas developed independently from, and sometimes unknown by, the Old World. The Mayan rubber ball, simple and brilliant, hasn't really been improved significantly since its invention perhaps ten thousand years ago. Dazzled as we are by the products of more recent genius, and dazzling as they indeed are, we too easily regard our ancestors as inferior and their achievements as obvious. The opposite is really the case: The most difficult and urgent problems were solved first, and these solutions still stand among humankind's greatest achievements.

Primary Colors

Our ancestors could paint a spectrum with natural pigments. Some colors and their ancient sources:

Red: Madder, a wild herb common in the Old World

Yellow: Saffron, also a food additive, and henna, which grows wild in arid climates

Blue: Indigo from various plants and marine snails, and lapis lazuli, the source of the pigment ultramarine

Purple: Tyrian purple, obtained from a common Mediterranean sea snail

Brown: Ocher, a common Old World mineral, the oldest paint pigment

1,000,000 BC

1 million BC *Homo erectus* lash stone knives to poles and learn to fling them effectively. Impact is instant, profound: the spear transforms the hunt and improves nutrition. Recent paleoanthropology suggests spears inhibited violence among tribes by greatly improving defensive capability, thereby discouraging attacks.

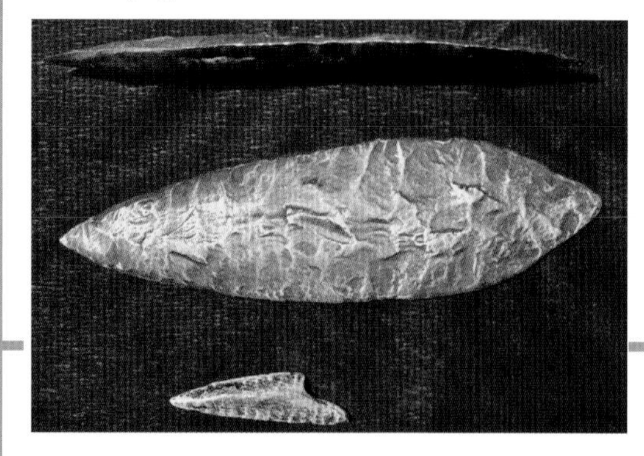

By 350,000 BC Predecessors of modern humans grind rocks to extract minerals, which when mixed with water produce the first paint. Earliest pigments made from ocher, iron oxide, other colored minerals used for body painting; earliest surviving cave drawings (such as shown) date from 40,000 BC, though the first ones could well have been made long before that.

By 75,000 BC Moving beyond the feathers and antlers *Homo erectus* likely wore as adornments, Neanderthals perforate mollusk shells to make beads and use them for necklaces: the first jewelry. Traces of ocher (shown) found near perforated shells in South African cave indicate they were painted as well.

50,000 BC Pioneers from Southeast Asia lash planks together to form the first boat; then discover and settle New Guinea. Around same time, ancients drift or perhaps row (with the invention of the oar) from island to island in Indonesia and other parts of Southeast Asia, navigating without charts or compass. Shown, a Greek trireme from the 5th century BC.

41,000 BC Discovered in Slovenia, the bone of a cave bear appears to have been intentionally altered by a Neanderthal, creating the first musical instrument. Other early instruments appear to be gourds or bones, played as they were or modified similarly to the Slovenia find.

50,000 BC

43,000 BC Mining begins when ancients in Swaziland dig to extract ocher for painting; first copper mines date from about 4500 BC, and flint mining, for material (shown) used in tools and weapons, begins about 4000 BC.

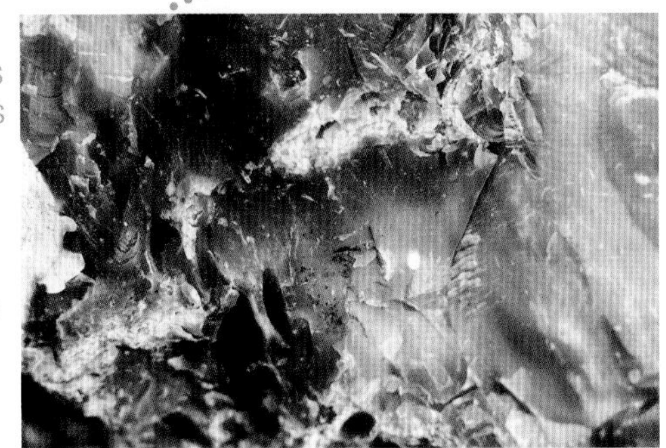

34,000 BC Neanderthals and *Homo sapiens* jostle for resources and primacy in Eurasia. *Homo sapiens* create a shelter using mammoth bones and wood, covering the frame in animal hides: the first house. Clustered houses lead to first villages, foster cooperation, encourage social evolution.

30,000 BC Mesolithic (Middle Stone Age) innovators lash antler to hand ax, creating the first true ax. The hand ax was known and used by ancestors of the human species, but the long handle allows for chopping of trees and eventually clears the way for agriculture. Romans later create blade with "eye" for insertion of handle, greatly increasing wood-clearing efficiency.

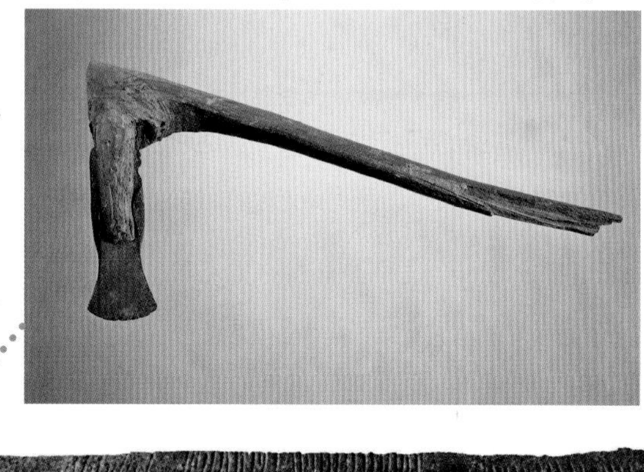

Adaptation vs. Invention

Picking a starting place for the history of invention is not easy, because it's not clear where nature leaves off and the human mind picks up. For example, apes and birds use sticks as tools to get at food they can't reach, so it would be inaccurate to say humans "invented" the tool. But they did invent the ax. Similarly, early humans are known to have tended natural-occurring fires and wrapped themselves in animal skins, but millennia went by before they developed a method to create their own fire or make what we would call clothing. Probably the real dawn of invention is best defined as that moment when people became conscious that they could find new ways to address their problems and started looking for those new ways deliberately. Artifacts show the pace of this process picked up considerably after plants were domesticated about 12,000 years ago.

30,000 BC

19,000 BC "Ishango Bone," found in Congo in 1950, suggests early invention of mathematics. Could also be an astronomical marker, but regardless of specific use, provides 1st example of information storage technology.

17,000 BC Paleolithic genius fashions carved wood tube—what came to be called an atlatl—that serves as a long hand-held launching cup for a hunter's spear, effectively increasing length of thrower's arm. Increases velocity of the spear, enabling larger animals to be brought down with same missile.

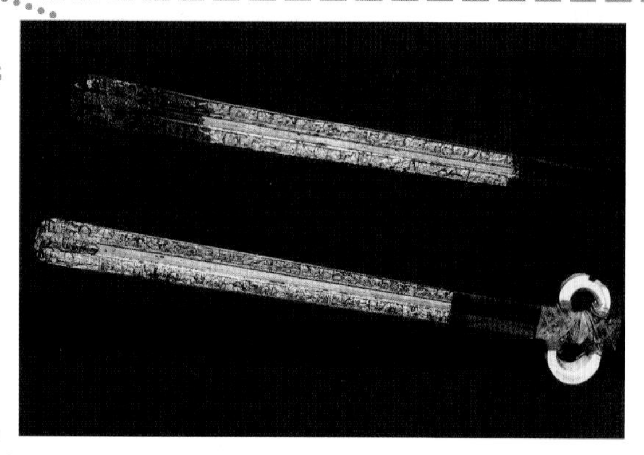

16,000 BC Process of baking clay to form pots and bowls invented in Japan, and independently soon thereafter in China and Russia. Pottery allows cooking of new foods, storage and carriage of water and other commodities over greater distances, and a host of new options for decorating and building.

13,000 BC Magdalenian (Paleolithic French) people invent harpoon by attaching hook, then hooks, onto spears. First harpoons seemingly used to hunt land animals, but the harpoon rapidly becomes the preferred method of catching fish, especially large ones.

12,000 BC

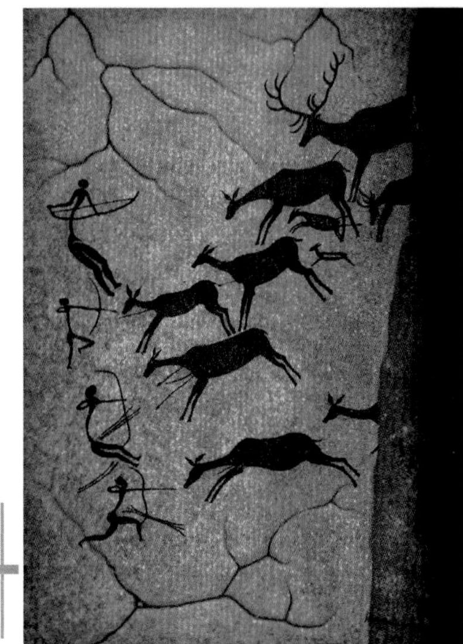

By 12,000 BC First harnessing of potential energy appears as hunters in the Middle East string flexible wood to fire short spears: the bow and arrow. Some scientists consider this the first machine. New firepower revolutionizes not just the hunt, but warfare.

Catching Fire

Pattern of pits and hearths reveals *Homo erectus* and other human ancestors may have controlled naturally occurring fire, carried embers to start fire for heat, cooking, and primitive agriculture more than a million years ago. Momentous shift comes in 12,000 BC when technology is invented for intentionally creating fire by turning a stick rapidly against a board and applying tinder to the glowing wood. Earliest surviving vestiges of intentional fire making are flint and metal strikers from c. 10,000 BC. Intentional fire provides first reliable artificial heat and light source, permits expansion of species to colder climates.

c. 10,000 BC Intentional sowing and cultivation of edible plants along with domestication of animals transforms pattern of human life from nomadic hunter-gatherer existence to settled villages; encourages development of art, religion, division of labor and specialization; and frees more time to invent and experiment. Developed about the same time in various parts of the world. Best evidence for first use in modern-day Jordan, Syria, Iraq, Iran. First crops: wheat, barley. Earliest livestock: goats, sheep.

9000 BC Indigenous peoples develop the basket, a light, strong container for transporting food and other goods, catching fish and, when coated with pitch, carrying water. Technique of weaving plant material important in evolution of fabric weaving, Wicker work adapted to use in furniture and other household items.

10,000 BC

Adobe

Hunter-gatherers worldwide develop brick making by intentionally allowing shaped mud to dry. Adobe, a mixture of mud and straw dried in rectangular molds, forms bricks so durable that structures last indefinitely in dry climates. Oldest known adobe structure is Bam citadel in Iran, though evidence also found throughout Near East and North American Southwest.

c. 9000 BC Mesopotamian heats up rocks containing copper and begins experimenting with hammering, melting, and bending the metal that runs out of the ore. End of the Stone Age and the beginning of the Chalcolithic: an era when copper tools were used alongside stone implements. Oldest example of metalworking is copper ingot from Iraq. Makes infinite number of manufactured items possible and leads way to invention of alloys and the Bronze Age.

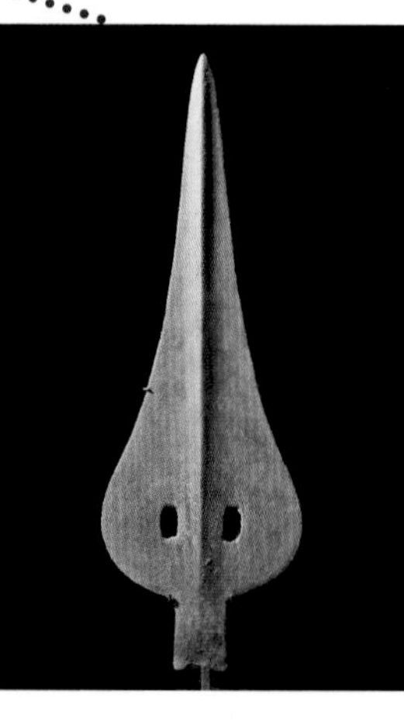

Cradle of Invention

When agriculture—a fixed food supply—was invented, so was free time. As the first farmers waited for their crops to mature and drew down their reserves, they had time to think. It stands to reason that a quickening of innovation took place shortly thereafter. Archaeology backs up that logic, innovations such as writing, money, shipping and sailing, along with myriad other smaller advances, all appear rapidly after the invention of agriculture. Many of the breakthroughs occur in Mesopotamia, the land between the Tigris and Euphrates rivers in modern Iraq, home to the Sumerians, Babylonians, Hittites, Assyrians, and Persians.

8300 BC Neolithic Britons, hunter-gatherers, lay the remains of 70 to 100 members of their community side by side in cave over 2 centuries—the first known cemetery. Later examples around Europe c. 4500 BC indicate growing sophistication of social structure.

7500 BC Ancient Turks polish obsidian (glassy hardened lava) to make the first looking glass. Later, Egyptians (2900 BC) and Chinese (1500 BC) invent methods to polish copper and bronze so finely, they produce the first man-made mirrors.

Who Invented Whom?

Many anthropologists view religion as an invention, born of a psychological need to be nurtured, to support a purpose for existence, to gain security of guardian spirits, or to explain mysterious phenomena. Many faithful believe their religion is delivered by a god or gods, not invented or discovered, yet consider other faiths to be fiction, i.e., an invented or at least misapprehended version of mystic reality.

7500 BC

c. 8000 BC Natufian tribe in Palestine invent modified ax with curved handle, and a curved blade made from flint—the sickle. Improves and speeds up harvesting wheat and other grains.

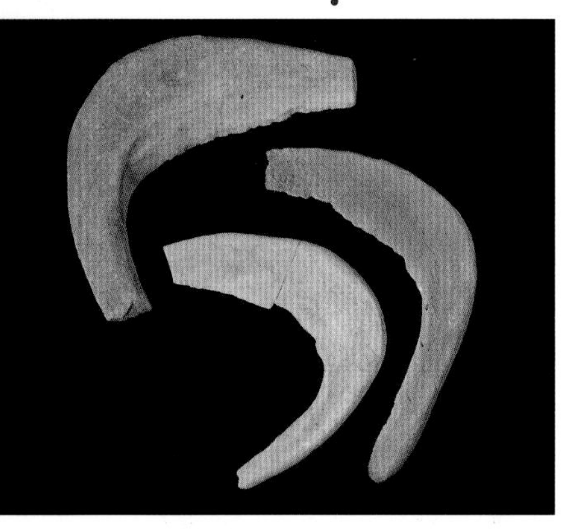

c. 7000 BC Pioneer cartographer in Catalhoyuk, Anatolia (modern Turkey) attempts to accurately depict layout of this busy trading center as if viewed from above—and carvings on a cave wall the earliest known example of a terrestrial map. Babylonians use clay tablets c. 2400 BC (seen here); much later in Greece, Eratosthenes develops latitude, longitude lines and attempts first realistic world map.

The City: A New Form of Living

Along with writing, city living defines "civilization." The example used by most scholars is Jericho, whose walls indicate the oldest purpose of city formation: safety. Cities that arose between 7000 BC and 3000 BC across the Near East were more than just larger walled villages; their arrangement heralded the dawn of human specialization and the rise of religion. Most early cities placed the temple at center, surrounded by homes and offices of professionals (priests, physicians). Workmen and the poor lived nearest the city walls. Rome estimated to have a population of 900,000 at its height of 50 AD.

6500 BC Inhabitants of the Near East adopt a basket weaving technique for the production of cloth fabric using flax (shown), a tough fibrous plant soon domesticated for use in weaving. Egyptian plagues in 3000 BC, which are mentioned in the Bible, are devastating primarily or their impact on the flax harvest, as modern archaeobotany now shows. Other cultures develop use of cotton, wool from various animals (sheep, camel, llama).

c. 6000 BC Assyrians construct an artificial waterway—a canal—in Nineveh (now Mosul, Iraq) to bring fresh water closer to fields and villages. Their potential for boat transportation is quickly recognized; canals serve as the super-highways of goods and commerce. Egyptians build Suez Canal along the course of the present one about 1500 BC.

6500 BC

c. 6000 BC Sumerians place barley in a dark container and allow it to ferment, creating the world's first beer. First grape vineyard appears around the same time (c. 5000 BC) in eastern Mesopotamia; unknown whether beer or wine, the product of fermented grapes, is the world's first alcohol. Later, Babylonians make a sort of portable beer by soaking sprouted-barley cakes in water; resulting beer is called "boozah," giving rise to the word now commonly used to describe all alcoholic beverages.

Holding Back the Water

History of dams begins in Babylonia (modern Iraq); some still hold water. Series of new uses for the dam can each be considered separate inventions: first use likely to create drinking water supply, then irrigation pools; important side benefit was creation of recreational pools. ventually, stored energy of dammed water is rigged to drive machinery, bringing about the advent of water power. Dikes use same technology, to keep water away from low-lying countries. First masonry dam was built c. 2700 BC across Nile. Oldest known dam in China dates from before 2200 BC.

First Worst Case

Sumer, the world's first civilization, had built by 3000 BC an immense irrigation system. But overexploitation of irrigation technology led to a disastrous accumulation of salt deposits, reduction in crop yields, and contributed to the eventual downfall of the empire. Sumer's collapse provides one of the first examples—but not the last—of the enormous negative consequences of ingenuity.

c. 5000 BC Egyptians invent the balance scale, a device that can assess the difference in weight between two items. Signifies first attempt at objective measurement of the physical world. Two millennia pass before notion of standards introduced. Fixed objects are assigned to serve as reference points to measure weight; first unit is the Egyptian kite (approximately 10 grams).

Play That Tune

Musical instruments evolved from primeval drums and pipes to stringed instruments to horns. Stringed instruments capable of different notes were created in Sumer about 4500 BC, marked by the invention of the harp. King Tutankhamen's tomb contained hunting horns of silver and bronze/gold dating from about 1340 BC, the oldest surviving metal wind instruments. By curving the tube, the French horn allowed for more notes to be played, and early versions of this instrument appear during the Middle Ages, along with reed instruments. (Keys and stops on horns were not seen until the 19th century with the invention of the trumpet.) Middle Ages also brought new methods of creating sound with strings: the dulcimer is played with hammers, and a bow is scraped across strings of rebec and vielle, forerunners of the violin and related instruments.

5500 BC Triangular "scratch plow" in Africa, Eurasia, Near East, and Middle East becomes synonymous with invention of agriculture and animal husbandry. A tree branch and, later, a blade would be hitched to a person or animal to create furrows; crisscross pattern used to create square fields. Material for blade evolved from wood to metal. Heavy plow, invented to deal with Europe's much wetter, heavier soil, arrives on scene much later (1 AD) and revolutionizes peasant life by expanding peasant farms, allowing cultivation of richer soil.

5500 BC Egyptians increase the size and productivity of their fields by digging the first irrigation channels from the Nile. Another Egyptian breakthrough from the same period: the shadoof, a bucket on a counterweighted pole that hoists water up to fields from river channel.

By 4600 BC Chinese invent knoblike clothes fasteners—buttons—made of pottery with holes for attachment. Greeks and Romans also use buttons, likely independently reinvented by Greeks; use not widespread until advent of fitted clothing in 12th c. AD; no evidence of buttonhole until that time (i.e., earlier buttons were fastened with loops).

3500 BC Sumerian and Egyptian jewelers independently establish how to join two pieces of metal by inventing soldering process: a metal with a lower melting point than the one being worked is dripped precisely onto the pieces to be joined, mixed with "flux," or plant resin, to make the metal bond well as it hardens.

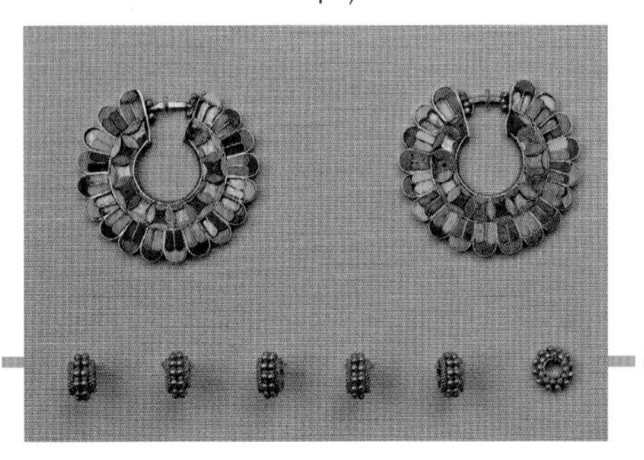

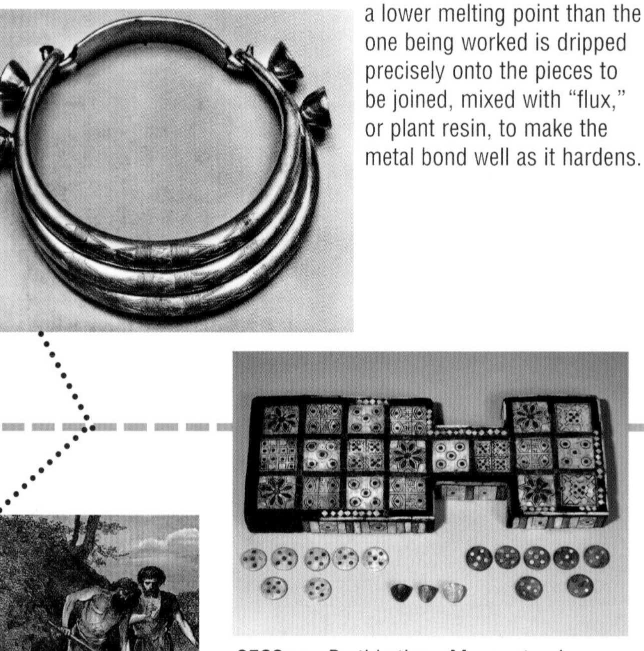

3500 BC

3500 BC Bronze Age begins when inhabitants of the Middle East melt and mix copper and tin, inventing not just bronze, but the whole realm of metal alloys. Unlike the copper that had been used in preceding millennia, bronze is hard, strong yet malleable, and will hold an edge nicely. Introduction of this first synthetic material transforms weaponry, tools, appliances and generates profound increase in quality of life and military effectiveness. The bronze arms race begins.

3500 BC By this time, Mesopotamians and Egyptians develop a new kind of amusement: the board game. Most ancient games (i.e., the Royal Game of Ur pictured here) are race games not unlike backgammon, controlled by dice, often made from the joint bones of livestock.

Inventing Time

The cycles of nature—day and night, the seasons, the phases of the moon, the movement of heavenly bodies through the sky—have always been part of human existence. Babylonians invented the calendar with 354-day year (12 lunar cycles), about 3000 BC, and later added months in some years to keep up with solar cycle. Babylonians also created a 7-day week and named the days after sun, moon, etc. Egyptians may have invented 365-day solar calendar c. 3500 BC, but not in regular use until about 1000 BC. Chinese lunar calendar dates from 2637 BC; Hebrew lunar calendar, from biblical times (c. 3500 BC).

3500 BC
Egyptians and, about the same time, Chinese increase versatility and strength of wood as building material by gluing thin sheets, one layer at cross-grain to the next, to form the first plywood.

3500–3000 BC Mesopotamians replace pack animals with draft animals, build wheelless sleds and hitch them to oxen. Soon, wheels are added to make the first cart. During same period, yoke is invented to harness, literally, the hauling power of multiple oxen.

The Wheel

The most common misperception about the history of invention has to be that the wheel is incredibly ancient—along with fire, the starting point of human ingenuity. Fire qualifies, but the wheel is nowhere close, apparently invented thousands of years after the plow, map, and beer. The first wheels were planks fastened together and then carved into a round shape; earliest evidence is from Sumer, with vestiges of first carts dating to 2600 BC. True origin of the wheel, and the reason it was so late in appearing, is still a mystery.

3500 BC Sumerians realize sounds, not just objects, can be represented in pictures, and develop semistandard code of symbols for sounds. Not only is writing invented, but human expression, information sharing, and record keeping are revolutionized. History begins. Sumerian cuneiform acts as basis of all alphabets; Egyptians use hieroglyphs; Chinese, pictographs.

c. 3000 BC Egyptians invent a new, durable, lightweight writing surface using stalks of papyrus plant to make thin, flexible sheets. Papyrus exported from Alexandria is dominant writing medium for thousands of years in Egypt and the Near East. Vellum and parchment (stretched thin animal hides) are other options developed before the coming of paper.

By 3000 BC Assyrians and Egyptians begin assembling furniture from wood—vestiges of chairs, tables, and stools survive from the Assyrians, whose records indicate the use of 17 kinds of wood in their furniture. Inventions behind the invention: innovative woodworking tools such as chisel, plane, and vise. Carpents are heavy wooden carts that lend their name to a profession.

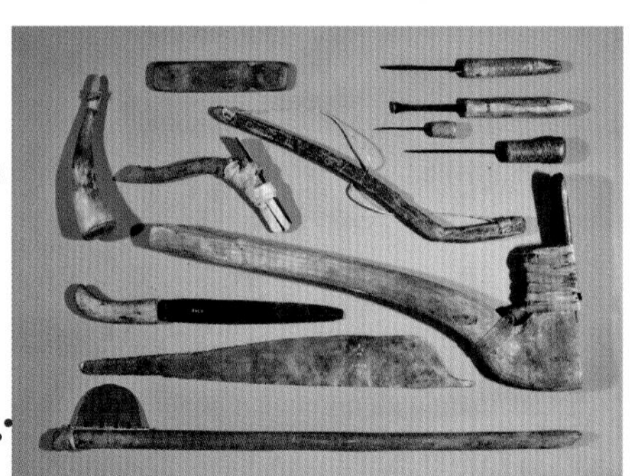

New Horizons

Sailing and navigation were not so much invented as learned. Ingenuity lay in recognizing and taking advantage of wind, weather, currents, landmarks, especially to exploit productive trade routes. Leaders in this innovation and experimentation: the Phoenicians, who sailed from coastal Syria and Lebanon c. 2800 BC, and the Mycenaeans, who traded with Egyptians, Greeks, and Phoenicians from their home on Crete in 1400–1200 BC.

c. 3000 BC Egyptians use the wheel to greatly improve the process of making pottery. Some archaeologists argue the potter's wheel, called a "kick-wheel," inspired the wheel for vehicles.

3000 BC

c. 3000 BC Egyptians affix mast and sail to river vessel; power of wind harnessed for first time, though millennia will pass before triangular sails permit mastery of sailing in the modern sense.

3000 BC Having invented the wheel and cart, Mesopotamians create first intentionally cleared and maintained roadways. Later, Greeks shy away from extensive road building, believing it disrupts activities of the gods. Romans, unconstrained by this belief, improve roadwork to such a great extent and high quality, some parts of their network are still in use.

3000 BC Chinese develop silk making by harvesting thread from silkworms and, soon, organize its production by cultivating both worms and the mulberry bushes on which they thrive.

Production of silk spreads from China to India, Korea, and Japan by 100 BC. In about 550 AD, Byzantine Emperor Justinian I (527–565) arranges to steal mulberry seeds and silkworm eggs, and introduces sericulture to Europe.

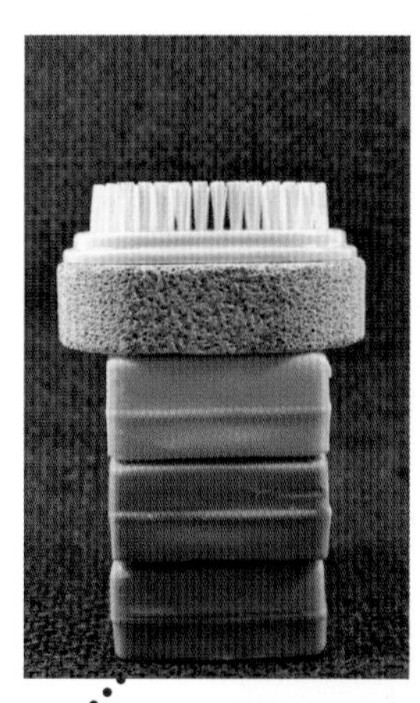

2800 BC Babylonians mix ash with goat fat, making first soap. Over the centuries, bathing goes in and out of favor depending on the time and place; soap making unknown in Europe until 1000 AD, when brought in by journeyers from Muslim lands.

c. 2500 BC Ancient Egyptian chemist melts sand and adds lime, producing the first glass. Numerous glass beads have been found from this period, though technology may have been understood well before; molds were used for creation of vessels and decorative objects until the invention of glassblowing 2,500 years later. Molded glass difficult and costly to produce, making it a rare luxury item for thousands of years. Blocks of glass produced in quantity at extensive glassworks in Alexandria (Cairo).

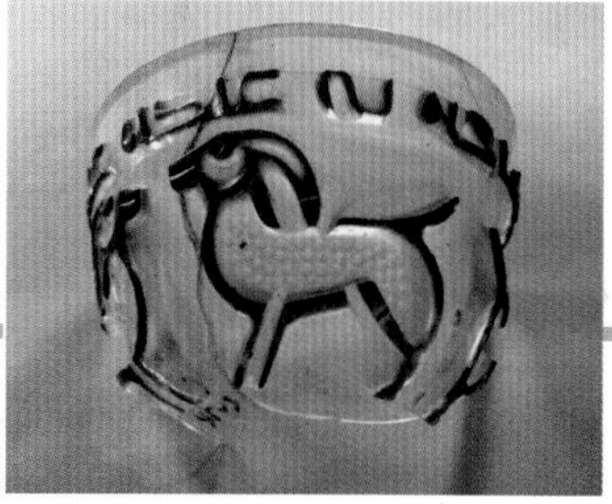

3000 BC Egyptian chemist blends gypsum and lime with sand: invents cement. Used in pyramids but not seen extensively until Romans refine formula and processing techniques to produce high-quality cement still holding many structures together after 2 millennia.

c. 2600 BC Egyptians first to carve stone into massive cylinders, starting a revolution in fine architecture with the first column. Columns elevate construction of both shelter and public spaces, allow building on grandiose scale as later exemplified by the Parthenon in Athens. Greeks develop distinct style for their columns (Sturdy Doric, elegant Ionic, slender Corinthian), which Romans codify in strictly followed written templates; column is symbol of Roman and later Western rule of law.

c. 2500 BC Egyptians engineer the arch, a new form of highly stable architecture in which shaped stones are stacked and held in place by scaffolding until the central keystone at the top locks them in place. Use in Egypt is mostly decorative and minor; Romans famously use it everywhere throughout their cities and civilization.

c. 2500 BC In agricultural records, Egyptians and Mesopotamians record the first use, therefore the invention of currency. They assign to precious metals specific, standard value based on quantity and use in trade for services or other goods

2500 BC Egyptian and Chinese chemists invent ink independently at what amounts to exactly the same moment on the historical record. Carbon dissolved in water with added plant gum will dry permanently on paper. India ink thicker, more complex substance; still used by artists. Coloring agents soon developed.

2500 BC

By 2500 BC Bronze workers hammer loops of metal together to form first chain; earliest examples found in tombs in Mesopotamia. First depictions, by Egyptians, show prisoners held fast. After invention, chains quickly supplant rope for security, restraint, and hauling.

2400 BC Fabric stretched over circular frame to protect user from sun—the parasol—depicted on monument to King Sargon of Akkad, Mesopotamia. Chinese first to develop umbrella, i.e., waterproof parasol, by oiling silk around 450 AD.

2000 BC Minoans on Crete invent and implement most aspects of modern sewage sanitation. Flush toilets (stream diverted to tank, water let in by lever that lowers flap) lead to clay pipes that carry wastes to septic tanks. Shown here, Roman toilet.

2000 BC Early Egyptians, likely Chinese as well, exploit tendency of animal fat to burn. They mold tallow (chopped and heated animal fat) to create first candle. Early versions are dim and smoky, but reed wicks become major advance in interior lighting. Experimentation with composition over eons reduces smoke and improves intensity and clarity of light. Beeswax also used, but reserved for the wealthy. Tallow candles carry advantage: they are edible!

By 1800 BC Likely invented in Babylon, the pulley employs two technologies, the rope and the wheel, to change the direction of force so that heavy loads can be lifted or pulled.

1500 BC

c. 1500 BC Egyptian priests tend lighthouse fires along the Nile coastline; build most famous lighthouse of antiquity (finished in about 280 BC) on the island of Pharos. Colossus of Rhodes uses mirrors to enhance warning at entrance to harbor; reportedly can be seen for 35 miles.

c. 1500 BC Phoenicians improve on prehistoric invention for watercraft propulsion—the paddle—by affixing it to a collar on the hull of a boat, inventing the oar. Advantage of lever power for boats, ships seen immediately; by time of Macedonians (c. 900–149 BC), ships have 1,800 rowers with 18 on each oar.

By 1600 BC Mesoamerican chemists experiment with sap from rubber plant, mix it with other substances such as the smoke of palm nuts to invent stretchy, bouncy material that renders fabric waterproof. Rubber later "discovered" by Columbus in 15th century AD.

1600 BC Northern Semitic people of Syria, Palestine develop a system of representing sounds with characters—the first fully formed alphabet—greatly simplifying written language. Earlier methods had used hundreds or thousands of symbols to represent whole syllables; Semitic alphabet saves time, effort, and writing material. Later improved by Phoenicians, who spread it through ancient world in their travels.

c. 1500 BC Egyptians introduce sandals (shown, modern Afghan sandals for males) specifically shaped for left and right feet. This is an improvement, even though ancient footwear already sophisticated, as evidenced by the 1991 Austrian discovery of "Otzi" the Ice Man (c. 3300 BC).

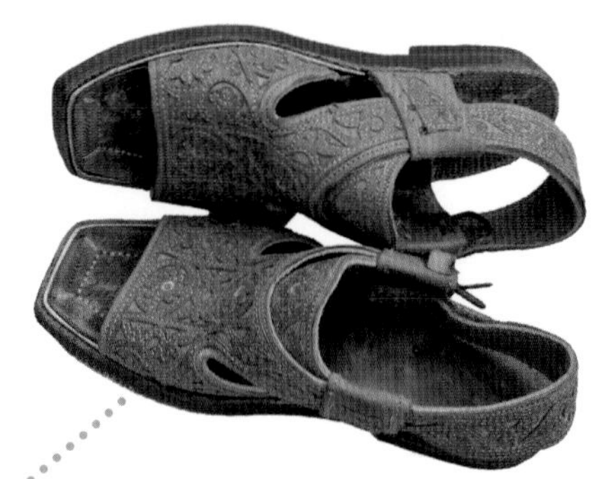

1600 BC

c. 1500 BC Spoked wheels improve speed and strength as innovations on the first wheel slowly emerge. Spoke begins soon after invention of wheel itself when engineers cut out holes in solid wheels to lighten them; experiments with this method eventually transform entire approach to wheel design and construction.

1400 BC Palestinians develop brass, a new alloy of two-thirds copper and one-third zinc. The new metal is light, strong, and resists tarnishing. Brass still favored today for musical instruments, buttons, bells, and clocks. Roman coins featuring Caligula and his 3 sisters (c. 12–41 AD) shown in image.

The Iron Age

Disruption in world supply of tin about 1200 BC forces an increase in smelting and refinement of iron, which is known since at least 4000 BC but is difficult to extract because of its high melting point and impurities. Mesopotamians realize its superiority to bronze, especially for weapons, and usher in the Iron Age. Just as bronze armor defeated stone weapons, now iron weapons, particularly the sword, lend military dominance to the leaders in a new arms race. Iron continues to be difficult to produce in quantity until 1350 AD and the invention of the blast furnace.

Water Clocks

We think of the sundial as the oldest time teller, but water was used before the sundial to mark passage of hours. The Chinese water clock consisted of a jug over a bowl; the bowl was graduated and showed time elapsed since flow started. American Indians put one bowl with a hole in its bottom into another full of water. As water trickles into the inner bowl, markings record the passage of time.

c. 650 BC King Ashurbanipal of Assyria (Iraq) conceives idea of the true library—a depository for learned works, systematically collected and cataloged—and sends henchmen to acquire tablets from across the ancient world. Earlier examples of the library, not as methodically assembled, existed in Babylon and Egyptian temples. By 330 BC the famous library in Alexandria, Egypt (shown), contain 700,000 volumes (tablets and papyrus rolls); in same year, Greeks invent the public library.

c. 800 BC Egyptian scientists mark tablet with full complement of hour indicators to make the 1st true sundial. Earlier, a stick or pyramid without markings might be used. Length of hours (i.e., daytime broken into equal parts) varied with season. Circa 300 BC Greeks introduce variable markings (equal-length hours).

c. 640 BC Gages, king of Lydia (southern coastal Turkey), advances long-standing practice of counting money in standard physical units (shells, feathers, beads) by ordering the production of metal discs specifically designated to serve as value tokens. The coin is born. Gages's successors continue the practice, including Croesus (561–546 BC), whose holdings in gold and coins gave rise to phrase "rich as Croesus." Monetization of ancient world rapidly follows.

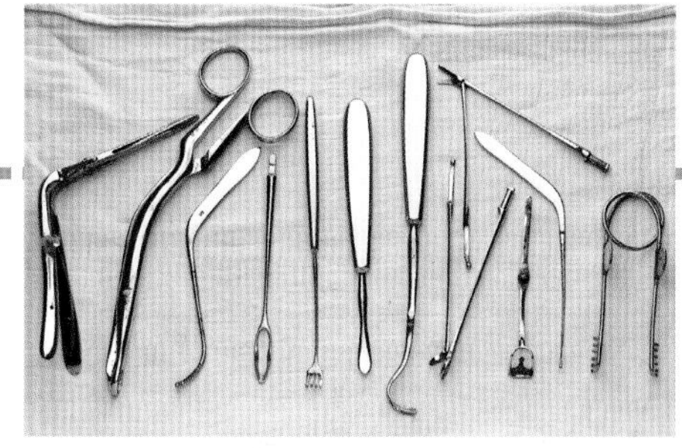

c. 600 BC Sushruta, considered by Indians to be the "father of surgery," invents rhinoplasty, a surgical technique for nose reconstruction that uses skin from the forehead. Still in use today, it marks the dawn of plastic surgery.

c. 600 BC Greeks invent purely mathematical method for investigating, testing, and proving geometric principles, introducing mathematical modeling approach to exploring spatial relationships so key in building the modern world. This mathematical method reaches its zenith with Euclid (c. 330–c. 260 BC), who is sometimes cited as the father of geometry. He can fairly be thought of as the patent holder on the mathematical method, even though practical geometry for surveying land, architecture, etc., had actually been used for millennia in Egypt, India, and Babylon.

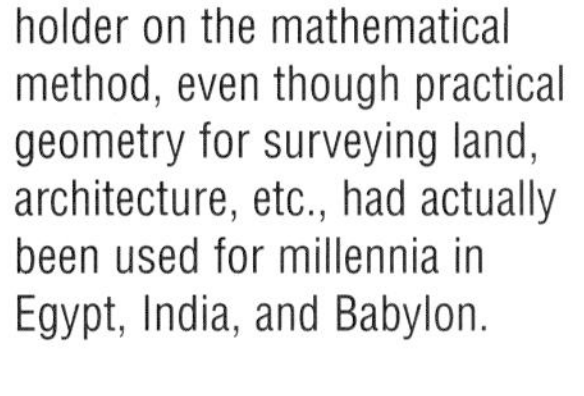

592 BC Greek historians report Anacharsis invents a major improvement on stone anchors already in use. He fashions a rod with a curved section on the bottom that catches the seabed, solving a vexing and longstanding (long-drifting, really) problem for early mariners.

500 BC

c. 500 BC To improve commerce with Asia Minor, Persians (Iranians) build first highway, i.e., a long-distance overland road. Chinese also construct highway network across their empire. The Roman highway system (shown) begins at roughly this time; 900 years later, at fall of empire, it comprises 50,000 miles of roadway.

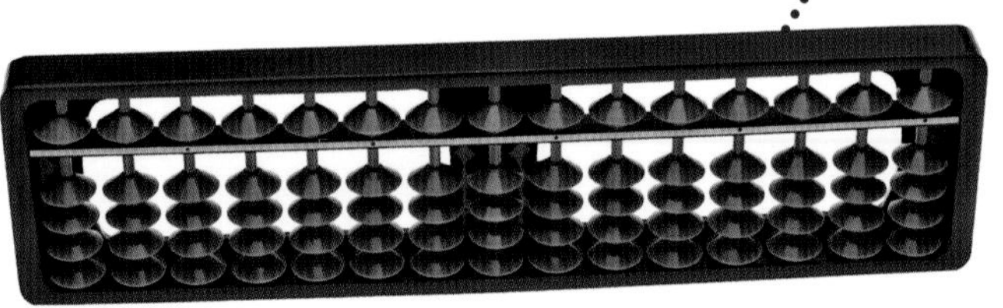

500 BC Chinese invent the abacus, which calculates so elegantly, it's still in daily use. Similar counting devices depicted in Babylonian art.

The First Name in Invention

Prior to about 600 BC, histories either don't record specific persons behind great inventions, or perhaps they incorrectly give credit to kings or emperors. Kings Ashurbanipal and Gages reliably do deserve credit for their innovations, the library and money. Euclid clearly did invent the mathematics of geometry (based on the detail and distinctness of the written record), and with his curved anchor, Anacharsis can safely be thought of as the 1st inventor in the common sense: the creator of a useful gadget.

c. 400 BC Greeks expand on existing idea of the ballista—predecessor of the crossbow that shoots arrows with elastic energy of stretched and twisted rope—by massively increasing its scale, thus inventing the catapult. It provides a phenomenal advantage in siege warfare; catapult principle used today to launch warplanes from aircraft carriers.

c. 300 BC Romans mix ice with fruit and honey, inventing first "Italian ices." A renaissance, so to speak, occurs later in the 14th century AD when Marco Polo brings recipes for ices back to Italy from Asia.

c. 300 BC Romans invent chain mail, a metal fabric composed of small interlocked metal rings, heavy but capable of blunting the damage from swords. Body armor slowly grows more expansive and elaborate until the 18th century, when firearms are made smaller, more accurate, and more deadly, making heavy body armor obsolete.

300 BC

312 BC Romans construct first system of water transport via aqueducts into lead pipes, leading to faucets in some private homes and, by 200 AD, hot and cold running water. Aqueduct had been first developed by Assyrians, and Babylonians had drains and some pipes, but Roman invention of modern piping leads to elaborate public baths and extensive sewer system to carry off wastes.

c. 300 BC Chinese military engineers attach a crossbeam to a hand bow and equip it with a latch and trigger. The crossbow provides an immense increase in firepower and range. Later improvements, especially after the invention of steel, make it the preferred weapon of armies even some time into the era of firearms.

c. 250 BC Archimedes posits that the compound pulley multiplies a pulling force proportionate to the number of pulleys hitched together. He allegedly demonstrates value of his invention by hauling a ship onto land by himself; his "block and tackle" method still used worldwide. Among earliest applications: moveable props for special effects in Greek theater.

c. 240 BC Eratosthenes invents modern geography by positing five zones of latitude: arctic, antarctic, temperate, tropics. Also accurately estimates the tilt of the earth's axis, and the size of the earth itself, within 15 percent. Draws most accurate world map to date.

c. 100 BC Chinese notice that objects made of lodestone, a mineral of iron, tend to change direction by themselves, always shifting north. They make "spoon" of lodestone and pair it with a surface marked with directional points. A thousand years later, Chinese and others develop compass as a major advance over navigation by sun, stars, and wind.

250 BC

c. 240 BC Archimedes invents a machine using a hollow, pointed coil, which when placed in water and revolved, can lift water from low to high point. Archimedes's screw still used in some areas for water extraction and transport. Screw mechanics eventually applied to fastening devices, presses, corkscrews, and turn screws for adjustment of mechanical parts.

180 BC Breaking from practice of thousands of years, Greeks fold papyrus into sheets instead of rolling it, and bind it into "codex"—the first book. Information can now be accessed at any point in a document, instead of sequentially; documents no longer limited by length of roll. Chinese *Diamond Sutra* dates to 868 AD, qualifies as first printed book, made with engravings and probably clay block type.

87 BC Posidonius, brilliant Greek researcher and philosopher, is credited with inventing the Antikythera mechanism, the first "calculator." Finely crafted gears allow it to add, multiply, divide, and subtract. Used to measure lunar months, predict eclipses, and display the correct place of the sun and moon within the zodiac. Before 1901, when it was discovered in the remains of an Aegean shipwreck, scholars believed gear works weren't invented until after 1500 AD.

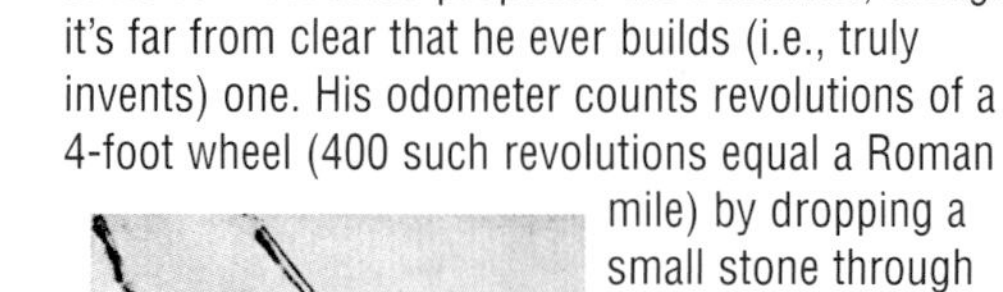

c. 25 BC Vitruvius proposes the odometer, though it's far from clear that he ever builds (i.e., truly invents) one. His odometer counts revolutions of a 4-foot wheel (400 such revolutions equal a Roman mile) by dropping a small stone through a perforated disc into a box for each mile traveled. In the 16th century, Leonardo da Vinci constructs a working odometer (seen here) based on the design of Vitruvius.

c. 0 AD Syrians greatly reduce difficulty of producing glass vessels by introducing glassblowing, or puffing air through a tube, at the end of which is a glob of molten glass (sand and lime). Glass vessels and objets d'art become much more widespread.

0 AD

c. 30 BC Vitruvius, a Roman engineer, describes his radical improvement on the horizontal waterwheel, known since 90 BC. In the upright version, little power is lost when the paddles on the wheel make their return trip to the water flow. First true waterwheel immediately adopted universally; production of mills rises exponentially. The "guts" of water mills and, much later, windmills, represent the arrival of complex machinery as an important part of human life.

c. 0 AD Chinese develop innovative method for heating: raised floor is constructed over an oven, creating the first central heating. At about the same time, Romans use columns and ducts like the system shown here, to carry heat under and into living chambers. Idea reputed to be that of Caius Sergius Orata, an oyster farmer trying to keep his stock warm.

100 AD

c. 100 AD Chinese mount a blade on the rear (stern) of a boat, making steering and sailing windward (into the wind) much easier. Rudder technology will not make its way to the West for a millennium.

118 Chinese inventor centers a single wheel underneath a cart, yielding the first wheelbarrow. Later, Europeans move wheel to front (shown here), increasing maneuverability and making lever power available when dumping load.

105 Cai Lun, Chinese minister of public works, comes up with an entirely new method for creating writing material. He chops up mulberry bark and soaks it into a pulp, spreads it on a porous screen, and lets it dry. This method remains the fundamentals of paper making.

200 Moving beyond the Greek pillar-and-slab design and African rope bridges, Romans experiment successfully with a bridge design so effective, it is still in use today: arched bridges spread load across entire structure instead of solely on pillars, making much heavier crossings safe and minimizing occurrence of collapses.

1 2 3 4 5
6 7 8 9 10

c. 500 Hindu scholars develop superior new method of counting and manipulating quantities: the decimal system. Massively improves economics and science because quantification becomes quicker and easier to standardize. Idea spreads to Arab world. By c. 750, when system spreads to Europe, numerals called “Arabic.”

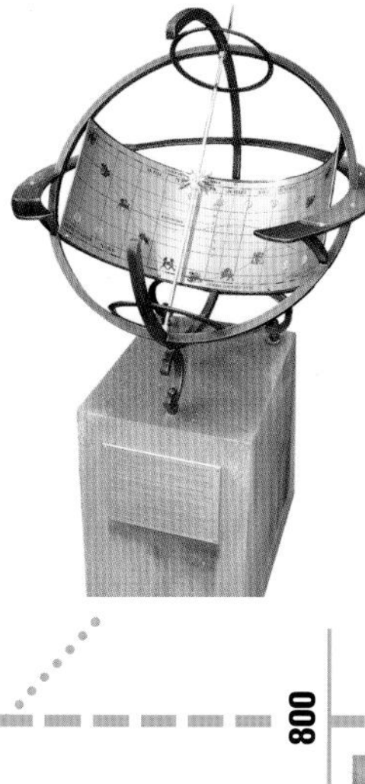

550 John Philoponus of Alexandria, Egypt, writes first known description of astrolabe, astronomical calculator that resembles a circular slide rule and protractor. Pointing one hand at celestial object and another at horizon, then reading marking on astrolabe, yields wealth of data about time of day, position of traveler in desert, even moment of next Muslim prayer; oldest known example is from 8th-century Baghdad.

810 Muhammad ibn Al-Khawarizmi, inventor of algebra and most respected of mathematicians, in Baghdad's palace called "House of Wisdom," argues before caliph and the royal mathematician that 0 should be considered a true number. After 2 days of debate, caliph agrees; 0 moves from being notational placeholder (i.e., to distinguish between 86 and 806) to an actual number, making many sorts of complex maths possible, advancing science, commerce, engineering.

0

800

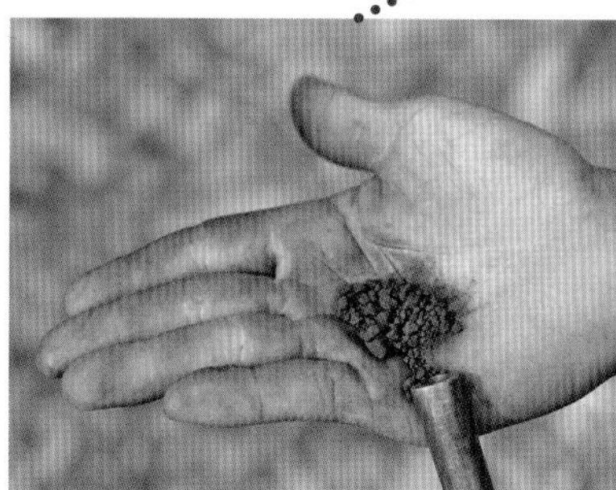

c. 800 Chinese alchemists mix sulfur, saltpeter (potassium nitrate), and carbon and generate first man-made explosion. Gunpowder is a low explosive, meaning it burns very quickly but does not "blow up" per se; its use lies in shooting projectiles when lit within a confined space. Finding the most effective tube to confine it and projectiles to launch it constitutes majority of weaponry science for more than 1,000 years. Forms the basis for development of high explosives such as TNT.

c. 824 Chinese invent the crank to operate a water-powered bellows, but technology of the crank not truly exploited till early 1400s in Europe by carpenters to drive tools.

856 Woodworkers in England invent the guild, a professional training and accreditation organization. Apprentices work under the eyes of master craftsmen for about 7 years, then become freelance journeymen; masters selected from journeymen and accredited to train apprentices. Invention of guild fosters transmission of knowledge and furthers innovation in many fields. The drapers' guild, immortalized by Rembrandt in 1662, shown here.

c. 800 West Asian sailors bring the lateen (the white sail shown here) from the Indian Ocean to the Mediterranean Sea. This triangular sail increases accuracy of vessel control, revolutionizes water travel and commerce.

983 Chiao Wei-yo, Chinese public works minister, has flash of brilliance about problem of moving ships and cargo from higher to lower sections of river. Orders construction of box with V-shaped doors facing upriver at each end; doors have flaps to let water out and in. By filling or draining, boats can be safely raised or lowered. Canal lock system eventually leads to 4,500 miles of canals in U.S. by end of 19th century. English lock shown here.

1000 After falling out of use for thousands of years, the fork reappears in Italian noble households; utility seems obvious, but widespread adoption problematic: Church bans use as immoral substitute for God-made implements, the fingers; Britons consider the fork too Continental.

1023 After several experiments in currency printing, including the use white stag hide, Chinese government begins printing paper money in quantity. Marks the beginning not just of popular use of paper currency, but of inflation, and soon, the invention of coun terfeiting.

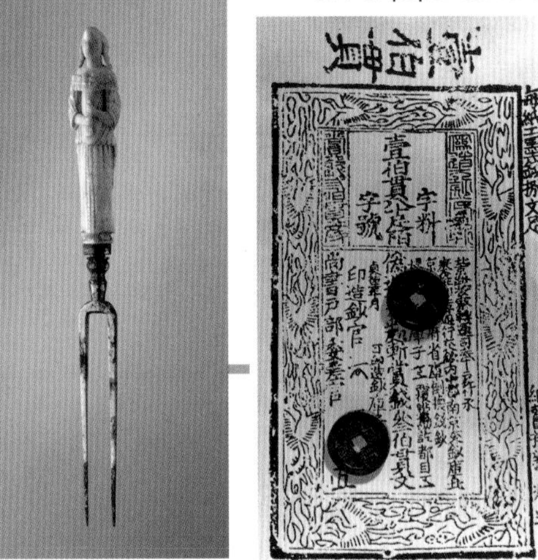

Not So Dark

Though the "Dark Ages" in Europe are marked by social disunion, and slow intellectual progress, glimmers of light are discernible in important agricultural innovations: three-field system of crop rotation; heavy plow for working tough, damp northern European earth; horseshoes and chokeless horse collar to make heavier plowing possible. All these lead to food surpluses and nuture a market economy, setting stage for Renaissance.

1041 Bi Sheng invents moveable type, using clay characters. Later systems use more durable material: wood (shown here) and eventually metal. Though Chinese do make some presses, page printing and bookmaking remains hindered by lack of workable press.

c. 1150 French introduce a major enhancement of the catapult called the trebuchet. It uses gigantic counterweights that, when released, swing up a lever arm loaded with heavy stones or, sometimes, the bodies of plague victims (an early form of biological warfare). Greater accuracy is significant advantage over previous designs.

1304 Arabs said to construct the first gun by packing gunpowder into a bamboo tube reinforced with metal; shoots arrows. Chinese may have shot some "fire lances" (rockets) from tubes much earlier.

c. 1100 Venetians develop coating of tin and mercury, creating the first glass mirror; Greeks may have known process, but lost it during the Dark Ages. In all phases —ancient obsidian, Greek, Venetian, and modern silvering process used today—the mirror becomes one of humankind's most important metaphors, for inner as well as outer self-reflection.

1260–1280 Flourishing of glassmaking and glassblowing in northern Italy leads to production of lenses and, by 1260, first spectacles. Exact inventor unknown: Italians Salvino degli Armato and Alessandro di Spina vie for the honor. Roger Bacon allegedly gives Pope Clement IV a pair about 1268.

Sparks Fly

After the invention of the cannon, the musket and handgun appeared. The cannon and firearms with their destructive projectiles made armor obsolete. But efficient ignition of gunpowder remained a problem; early fuses gave way to flintlock, in which a pressed trigger caused flint to strike an iron plate, spraying sparks. Problem eventually solved in the mid-1800s.

c. 1350 Water clocks, sundials unworkable or unreliable at sea, so sailors, most likely Italian, obtain barbell-shaped glass goblet and mount it on wooden frame, inventing the hourglass; significant advance in accuracy, utility of timepieces.

1300–1350 Arabs, Chinese, Europeans all develop metal tube that, packed with gunpowder, can deliver large and heavy ammunition hundreds of yards. Cannon, first of bronze, then of iron, reinvent warfare; earliest record of use in Europe is English against the Scots in 1327.

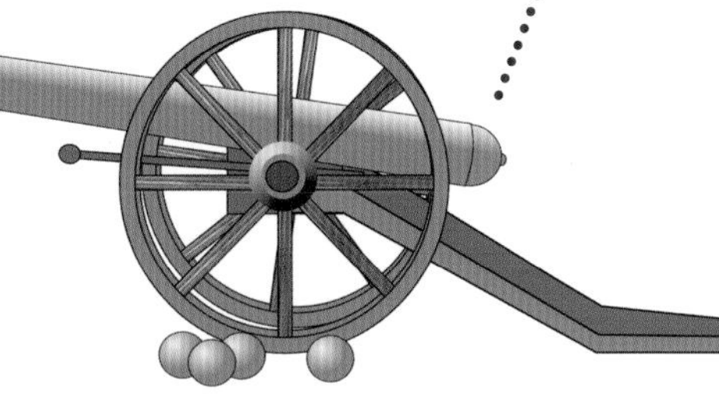

1350 Brothers at the Rievaulx Monastery in England, masters with iron, search for ways to produce more iron of higher quality. After long experimentation, they rig bellows to waterwheel to get fire super-hot; breakthrough comes when monks heat the air by circulation above furnace before it's pumped onto the flames. Blast furnace makes production of iron in large quantities feasible.

An Iron Framework

Monks at Rievaulx set stage for world transformation by making blast furnace viable. As furnace is copied and refined, iron can now be made in large quantities, and cheaply. Availability of cheap, strong building material, and ability to cast very large structural elements sets up conditions for modern urban life and Industrial Revolution.

1445–1455 Johannes Gutenberg develops a process for printing pages with moveable type cast from his own special alloy. Chinese moveable type and printing had had smaller impact because 80,000-character alphabet still made printing laborious; but Gutenberg's refinement and combination of existing processes utterly changes the world of communication and learning, and the world itself. Renaissance begins. Best indicator of scope of revolution: number of printed books in Europe went from nearly none in 1450 to several million only 50 years later.

1400

1410 Portuguese combine square sails with lateen (triangular) sail invented by West Asians, on 2- or 3-masted ship called the caravel (or carrack). It has no oarsmen, is light and highly maneuverable, especially good at sailing into the wind. Caravel opens way for Portuguese to explore Africa, Asia, Americas; new phase of world history begins.

Like Clockwork

Water clocks and sundials had long been known, but as the fundamentals of machinery (gears, levers, fasteners) become more widely known, mechanical timepiece becomes inevitable. Su Sung of China builds a clock with hands driven by a water wheel; next mechanical clock seen in historical record by Giovanni di Dondi. Clocks (1364) used weights; watches (1511) had to wait for improved spring technology.

From the Printing Press to Interchangeable Parts: 1447 to 1791

In the period between Johannes Gutenberg and Eli Whitney, the history of technology became a roll call of immortal names. In antiquity, credit for inventions went to civilizations and societies. Beginning with the Renaissance, invention and discovery became the province of individuals. This reflects a profound shift in thinking, just one of the repercussions of Gutenberg's printing press—an invention that arose from, and then fueled, the Renaissance. For the first time, learned people came to believe that individuals could question traditional wisdom dispensed from on high and take it upon themselves to search for new answers to old questions, new solutions to old problems, and new ways of choosing to view the world.

In the 1500s, scientists and explorers (and theologians) reinvented humanity's understanding of reality. In the 1600s, the scientific method was articulated, and science and invention were formalized in academies and societies. Phenomena such as magnetism, long regarded as a mysterious heavenly wonder, were rendered quantifiable, predictable, and understandable. In the 1700s, the science of both steam and electricity was developed, soon to transform all human activity. The high mathematics necessary to deal with the problems of power transmission, lens making, and a host of other technical issues was invented. And the crucial craft of making machine parts was advanced profoundly—an often-overlooked but totally essential part of creating the modern world.

But all of this began with Gutenberg's press, which cracked open a tomblike stillness in the intellectual life of the West, changing political, social, and technological life forever. These changes are summed up in the word *revolutionary*, coined out of the most profound and transformative scholarly achievement of the age: Copernicus's theory that the earth revolved around the sun instead of resting at the center of the universe.

The new sense of daring was most dramatically expressed in the era of global exploration with sailing ships and the European conquest of the New World. Here again, bold leaders were challenging ancient "truths" like the earth's flatness and changing humankind's relationship to the unknown: challenging and subduing it instead of hiding from it in parochial timidity.

The outstanding inventions of the age reflected this tremendous change in mind-set. To explore, question, investigate: that was the focus of many technological breakthroughs in these 350 years. From the telescope to the microscope, the atlas to the chronometer, much of the most significant innovations of these centuries had to do with the enterprise of converting the unknown into the known. Inventions to help people learn were just as significant as those to help them live.

In earlier epochs, no accurate, permanent record was kept as to who developed or perfected a useful invention. Many items arose anonymously and could be credited to the gods; now, with humankind more confident and the means of communication more powerful, specific individuals begin reaping, and indeed seeking, honor and attention for their achievements. This was not accidental: a new ethos of the individual had arisen in the 1500s, and inventors were far more willing to ask and get their due. And the printing press was making the written record of accomplishment much more extensive and precise.

So, by the threshold of the Industrial Age, we have the Argand lamp, the Franklin stove, the Pascaline calculator, the Mercator projection, and the Newtonian telescope. Not only did the number of patents explode, the whole system of patent issuance to protect the rights of individual inventors was professionalized, taken out of the realm of royal boon. And with patent rights came patent disputes, of course; this is the beginning of a long tradition of fierce and often unseemly quarrels between intellectual giants over who had an idea or invention first.

One of the most important features of this period (and this section of the timeline) is the fading, the disappearance really, of Asian invention, as China and the Middle East ceded their long-held dominance in technological development. Several theories compete, or combine, to explain this striking cessation. In China, the emperor Xuande decided to end China's great voyages of exploration in 1433, and a unified China withdrew into an isolationism that kept it from receiving and advancing the wonders of the Renaissance.

The Middle East did not yield to the West as dramatically. In fact, during the "Dark Ages," the light of learning was tended and strengthened by Arab civilizations such as the caliphates in what is now Iraq, where our number system, algebra, and the astrolabe had been invented. But after the glory years of Baghdad in about 800 AD, the importance of the Near and Middle East did dwindle compared to those accomplishments and seminal breakthroughs of ancient Egypt and Mesopotamia. Throughout the Muslim world, leaders and followers chose to spurn intellectual interaction with Western nations, their bitter infidel enemies in the Crusades.

But the cities and nations of Europe teemed with contact, competition, and intellectual controversy. The accompanying exchange of ideas produced a string of immortals (Leonardo, Bacon, Galileo, Hooke, Huygens, Newton) and thousands of lesser known but highly worthy inventors and scientists. Their work turned the Renaissance into the Scientific Revolution, and in time, the Industrial Revolution. By the end of the period, the focus was shifting fast to the New World, where the United States began to display a pragmatic genius fueled in part by its novel experiment with liberty. From the atom to the universe, from the smallest miracles of human anatomy to the teeming tracts of endless sea, new worlds of every kind were discovered and explored, and a wealth of innovation and progress resulted.

Leonardo da Vinci, Renaissance Man

Leonardo da Vinci personifies the Renaissance, not least for the way he expresses its unfettered and courageous new thinking in practical areas. Most of Leonardo's proposals are brought to fruition later, when technology catches up with his mental ability and foresight. Among Leonardo's many original ideas:

- Helicopter
- Parachute
- Armored tank
- Screw lathe
- Submarine

1450 Italian architect and "Renaissance man" Leon Battista Alberti constructs the first anemometer. Wind, pressing on metal disk, swings it along scale for readout of speed. Robert Hooke reinvents it in about 1700; familiar cup anemometer, such as seen here, not conceived until 19th century.

1498 First mention of a toothbrush, in a Chinese encyclopedia. Prior to this, twigs, rags, and rope used for teeth cleaning, and urine as dental antiseptic. Tooth brushing becomes widespread in America after WW II when returning soldiers continue daily hygiene drills.

1450

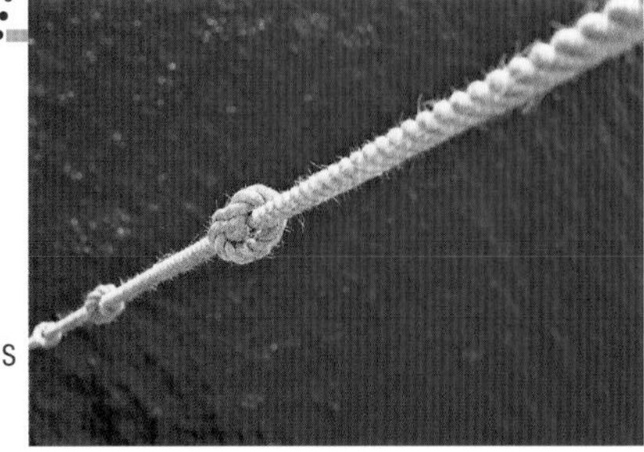

c. 1480 Leonardo da Vinci proposes ship's log in his notebooks, and the idea is implemented rapidly. Log is not a record book, but literally a log wound with a knotted rope; rope dragged out by a float as ship moves along and sailors count the knots to determine speed. This is why nautical speed is measured in knots.Timing accomplished first with human pulse, later with clocks. Also called chip log (the float was in the shape of a chip of wood)

1500
Peter Henlein of Nuremberg invents a portable clock—i.e., the watch—run by a spring instead of by weights. Henlein first gets credit for invention in 1511. Early watches not at all accurate: equipped with only an hour hand, they lose/gain a half hour per day. For first century, though much admired and sought after, they are mainly a beautiful novelty for the rich.

1510 Lucas van Leyden of Holland creates copper etching method to make finely detailed illustrations for printing press. Most of the illustrations we associate as highly accurate but "antique"—i.e., those by Henry Gray (*Gray's Anatomy*) and John James Audubon—are copperplate etchings, or engravings using the same kind of process with another metal. Mapmaking, scientific study are both significantly advanced by the fine renderings copper engraving makes possible.

1528 By 1500, Spaniards have harquebus, an inaccurate firearm called "handheld" but actually so heavy, it usually needs a supporting stand; they rely on S-shaped "serpentine" matchlock igniter that brings small smoldering flame to powder. Spain's next step is the musket, common by 1550. The musket, the first true rifle, uses much superior flintlock and is truly handheld. Beginning of the "musketeer" and modern warfare, and the end of the armored knight.

1513 Henry VIII's naval engineers design and build the first true gunboat, the *Mary Rose*. She and 4 others like her, with watertight gun ports and other innovations, launch new era in maritime warfare. *Great Harry*, pictured here, is built in 1514.

Printed Power

All through the 1500s, books, tracts, and pamphlets are produced at a prodigious rate; each one gives people more to think about. Knowledge, once the province only of men of religion with access to hand-lettered books, now available to the middle class, whose numbers are increasing as better technology raises the standard of living. Dissemination of information increases independent thought, questioning of accepted natural order. Printed accounts of the New World fuel this sentiment, reflected in religion in the Protestant Reformation; in science, Copernicus's revolutionary theory of the solar system.

1540 Valerius Cordus of Germany, pharmacist and botanist, develops method to synthesize "sweet oil of vitriol"—ether—by combining alcohol and sulfuric acid. Eventually, ether is used as world's 1st anesthetic for surgery. Shown, the Morton inhaler, devised in the mid-1800s for ether administration.

1565 Swiss scientist Conrad Gesner describes invention of the "lead" pencil —insertion of stick of fine Cumbrian graphite into wooden holder. About 200 years later, Nicolas Conté of France develops the mixture of graphite and clay still used in modern pencils.

1569 Great surge of exploration, recognition of spherical earth, and quest for new geographical knowledge leadsto a major problem: accurately plotting straight courses on a flat map. Belgian Gerardus Mercator creates map that conforms shape of the earth into rectangle, which distorts edges but leaves center accurate, and allows every line of longitude to meet every line of latitude at 90 degrees; easy to use and permits accurate straight-line course plotting. No alternative invented since has supplanted it. Shown, the Mercator globe of 1541.

Round the Globe

Though Gerardus Mercator is famous for flattening the earth, so to speak, with his map system that simplified navigation, he also makes important contributions in globe making, a respected but laborious art. He scraps the practice of engraving wood or metal spheres, and instead prints ovate sections that are glued into place, yielding a much faster globe-making process.

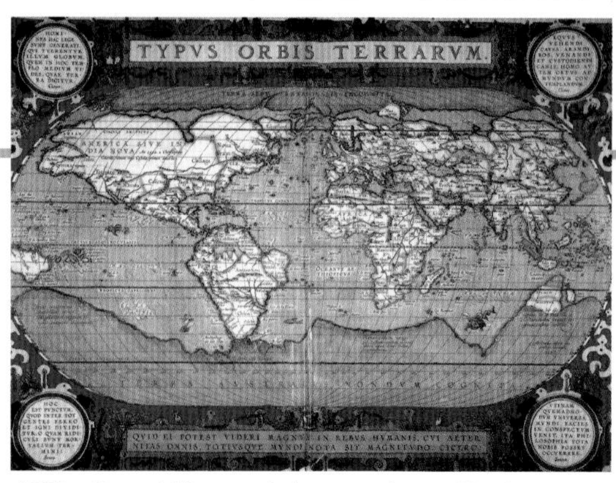

1570 One of Mercator's best students, Abraham Ortelius, thrives on selling maps but hears a complaint from customer: Navigation maps come in too many sizes and are illegible with their tiny type. Ortelius devises the first atlas, a collection of 53 maps that break up the world into parts, making each section easy for navigators to use.

1571 Modern surveying begins with the invention of the theodolite by Thomas Digges. He combines optical sights ("perspective glasses") with circular scales so that distances and angles can be read very accurately. A major advance in the accuracy of mapmaking during the golden age of cartography.

Measuring Nature

Adoption of the clock reflects a burgeoning change in attitude across Europe. Fearful faith in religious authorities' explanations for mysteries of nature begins to erode, and attitude toward nature begins to shift among learned people of upper class. Nature begins to be seen as explainable, definable, measurable. Push to accurately measure the hours a hallmark of this new attitude, as is quest for instruments like thermometer, barometer, anemometer.

1576 Legendary Japanese warlord and dictator Oda Nobunaga commissions the construction of at least six ships of war sheathed in iron, prefiguring the all-metal ironclads of the 19th century such as the famous *Monitor* and *Merrimack*, seen here.

1577 Tycho Brahe's increasing need for precision instruments to assist his astronomy work leads Jost Bürgi of Sweden to invent a clock with a minute hand.

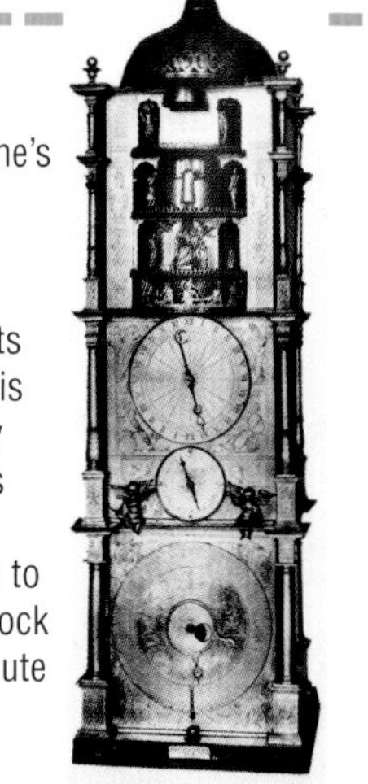

1582 Pope Gregory XIII declares the imposition of a new calendar throughout the Christian world, implementing invention of Aloysius Lilius, Claudius Clavius, and other mathematicians and astronomers. New rules for leap year keep calendar from slipping out of sync with seasons and important holidays like Easter. England, which separated from the Roman Catholic Church in 1534, does not adopt the new calendar until 1752.

1588 Englishman Timothy Bright revives art of shorthand by publishing *A Characterie: An Arte of Shorte, Swifte, and Secrete Writing by Character*. The practice of shorthand started with Greek historian Xenophon, who invented a Greek shorthand to capture the words of Socrates as early as 400 BC. The Roman orator Cicero, pictured here, had his words captured in *Notae Tironianae* by Marcus Tullius Tiro in 63 BC. But these forms were complex and died out. Today, even in age of digital voice recording, shorthand writers can still take dictation faster than anyone can speak.

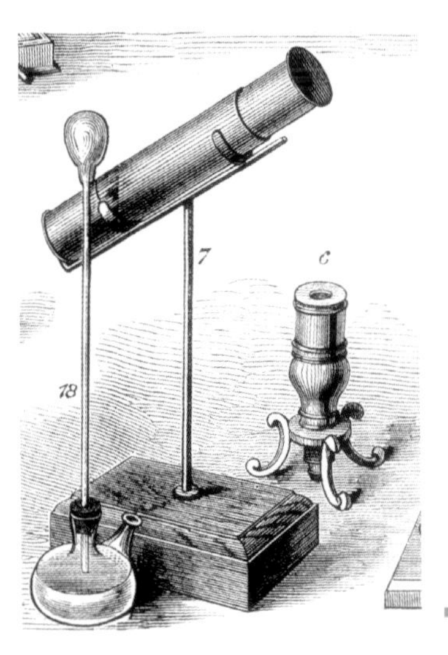

1590 Hans and Zacharias Janssen, father and son eyeglass makers in Holland, know convex (bulging) lenses magnify; place two together to create the microscope (shown, far right). In 1652 Antoine van Leeuwenhoek improves microscope and, along with Robert Hooke's engravings in his *Micrographia*, astound the world.

1589 Rev. William Lee invents automatic knitting machine; lore has it that his fiancée was more interested in knitting than in him, and that got him thinking. Rejected twice for patents by Queen Elizabeth I, the second time because the high quality of the machine's work made the queen fear the loss of many jobs in the hand-knitting industry.

1592 Galileo heats a glass bulb partially filled with water, out of which protrudes a thin tube, and the thermoscope is born. Water expands and rises with higher temperature, shrinks and falls with lower. Santorio Santorio, an Italian, first to turn thermoscope to thermometer by adding a scale in 1612. In 1714, German Daniel Gabriel Fahrenheit uses mercury for the first time and divides difference between freezing and boiling points by 180°. Fahrenheit chooses 32° for freezing, to be sure 0° is below coldest temperature he can produce in lab.

1597 Francis Bacon's *Essays and Meditations Sacrae* are published. Bacon is credited with "inventing" the scientific method—cycle of theory, test, revision of theory—around this time, though scientists in the Muslim world had already started an intellectual-scientific movement c. 1000 AD. But there can be no dispute about the impact of Bacon's provocative and persuasive declarations that scientists and scholars should discover truth through observation and testing, and reject traditional wisdom if it conflicts with the findings of impartial investigation. Also urges scientists to turn their talents to practical problem solving. Marks birth of modern scientific era in the Western world, which is typified, to use one important example, by the shift from alchemy to true chemistry. This also is the dawn of an epochal struggle between faith (religion) and reason (science).

Invent, Then Discover

For intellectual immortals of the 16th and 17th centuries, invention is an expression of scientific genius, often driven by a consuming quest to explore the natural world. Greats like Robert Hooke, Isaac Newton, and Galileo Galilei were capable of building the instruments to test the startling theories of the era, including their own. These are not just great thinkers, but brilliant craftsmen and technicians. Otto von Guericke invents the air pump, electrostatic generator, and a new type of barometer. Newton, studying forces binding the universe, invents both the reflecting telescope and calculus. Almost every important major discovery during this era begins with a brilliant invention to test a theory or conduct an experiment, by great thinkers who were also master technicians.

1600

1605 Johann Carolus of Strasbourg, Germany, who runs a news service selling expensive hand-written reports to subscribers, lowers prices and mass-produces; buys a printing press and begins publishing the first regular newspaper, the *Relation aller Fürnemmen und gedenckwürdigen Historien.*

1608 Hans Lippershey, an eyeglasses maker from Holland, arranges convex and concave lenses at the opposite ends of a tube and tinkers with their shapes and distance apart, until he produces astonishing results. As Galileo describes it to his brother-in-law, "a man two miles away may be distinctly seen" with this miracle device: the telescope. But authorities fail to see its potential and Lippershey is unable to win a patent. The next year Galileo confirms Copernicus's revolutionary heliocentric theory with a telescope of his own construction and discovers Saturn's rings, Jupiter's moons, and spots on the sun.

1612 Bedeviled by the time-consuming multiplication needed for navigation and science, Scottish mathematician John Napier simplifies calculations like .231987 x .529290. He publishes"A Description of the Marvelous Rule of Logarithms," explaining in effect how to turn any multiplication task into addition, and any division into subtraction. "Napier's bones," a logarithm calculator with rotating rods, makes these calculations even easier.

1623 Wilhelm Schickard of Germany, seeking to automate the calculation of logarithms, constructs "calculating clock," adding machine that employs bars, knobs, and cylinders for input. Numbers on the wheels display the output. Schickard's machine forgotten until the 20th century. Blaise Pascal, for whom the computer language is named, invents Pascaline calculator in 1642, and still receives credit for a breakthrough Schickard realized first.

1631 French engineer and mathematician Pierre Vernier invents a system of reading out the measurements made by scientific and surveying instruments. Vernier scale arranges large units along one line, smaller units on a parallel line, enabling quick, very precise measurements demanded by a new era of more rigorous scientific investigation.

1620
Submarine first appears when Cornelis Van Drebbel of Holland builds a working model from a 1578 design. Essentially a leather-covered rowboat, the oars stick through watertight seals in a waterproof leather cover, and oarsmen breathe through tubes attached to floats on the surface. British navy uninterested in the craft; Americans first to use submarine weaponry, when the *Turtle*, a one-man sub, unsuccessfully attacks a British ship.

Inventing the Patent

Before the advent of British trade guilds in the 1400s, inventors had no protection from theft of their ideas. (First patent law: Venice, Italy, 1474.) British monarchs singularly granted or denied claims and established monopolies until 1623 when the British Statute of Monopolies allowed inventors up to 14 years of exclusive rights to their inventions. High patent fees kept profits within upper classes. After 1700 the Crown granted patents based on nature and use of invention, leading to modern English patent law, which was far more egalitarian than Statute of Monopolies.

1632 Like others in his field, William Oughtred, a British mathematician, is thrilled with the invention of logarithm tables for simplifying multiplication and division. He spends a year perfecting his obsession: the slide rule, unveiled in 1621. By sliding an indicator across a fixed scale of quantities, users quickly find appropriate logarithm. With refinements, the slide rule spends 350 years as the indispensable calculator no mathematician is without.

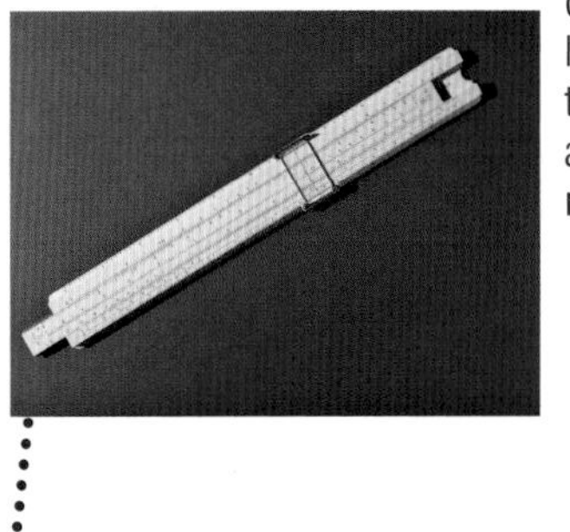

1643 Evangelist Torricelli, who carried on Galileo's work after the great scientist's death, inverts a mercury-filled tube in a bowl of mercury, notices some of the mercury stays in the tube, and concludes atmospheric pressure holds up the mercury. After levels rise and fall over a period of days, Torricelli realizes he's invented the barometer. Atmospheric pressure still given in inches of mercury.

Showman Scientist

Otto von Guericke turns heads not just for his science, but his showmanship too. His demonstrations of the vacuums he could create were as memorable as he could make them. First, he puts a bell in a glass container, showing it makes no sound if rung in a vacuum. Animals are sacrificed to show the fatal effects of a vacuum. He demonstrates the vacuum's strength with "Magdeburg sphere," two metal hemispheres with valves for evacuating air. With Holy Roman Emperor Ferdinand III in attendance, Guericke hitches two teams of horses to the spheres; when vacuum is created, horses can't pull spheres apart. Guericke turns a valve, the vacuum is released, and spheres separate easily.

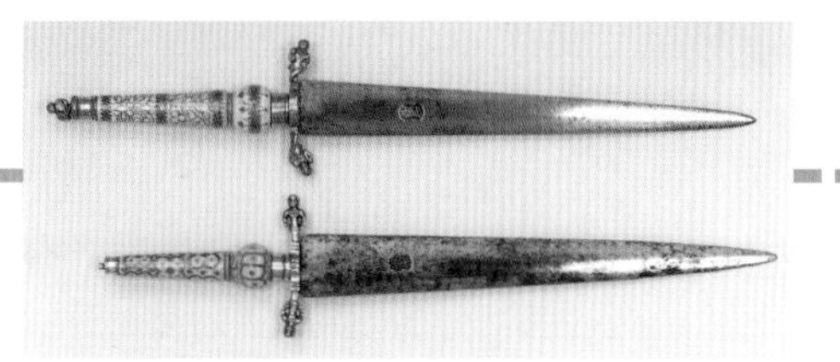

1647 Bayonet makes its first appearance in Bayonne, France, when peasants attach spikes to their muskets after they run out of ammunition. Bayonet allows foot soldiers to protect themselves while reloading, but musketeers must choose between using the gun or the blade, which fits inside gun barrel as those shown here. With invention of Scottish general Hugh McKay's ring bayonet in 1688, which fits around gun barrel, the weapon alters basic field tactics and the course of battles.

1647–1654 Otto von Guericke, mayor of Magdeburg, Germany, invents a piston air pump that allows the user to evacuate the air from sealed containers and disproves Aristotle's contention that creation of a vacuum is impossible.

The Priority Problem

The Leibniz-Newton controversy provides just one example of a new phenomenon linked to the coming of the celebrity scientist. With the Renaissance, individual achievement, especially scientific achievement, begins to be recognized and rewarded as never before. This leads to bitter and complex disputes in fields from astronomy to zoology as geniuses spend time feuding over credit. A (very!) condensed list of combatants: Galileo vs. Scheiner over who discovered sunspots; Huygens vs. Hooke over who invented the balance spring for watches; Newton vs. Hooke over several discoveries (reflecting telescope, laws of gravitation). These fights begin a scientific tradition still very much alive, reflected in disputes such as which group should get the most credit for mapping the human genome.

1650

1657 Inspired by Galileo's work with the physics of pendulums, Christiaan Huygens constructs the first pendulum-regulated clock, capitalizing on the pendulum's consistent cycle regardless of the distance it travels. Result: enormous increase in timekeeping accuracy.

1660s Rival geniuses Robert Hooke of England and Dutchman Christiaan Huygens both invent the balance spring, thin coiled metallic strip in the back of analog clocks and watches, making them useful at sea, where pendulum-based clocks do not work.

1660 New British Royal Society marks beginning of the world's first enduring scientific society. Membership contingent on at least one new discovery; meeting of minds accelerates knowledge sharing, establishes modern pattern of peer review. French Academy of Sciences founded 1666.

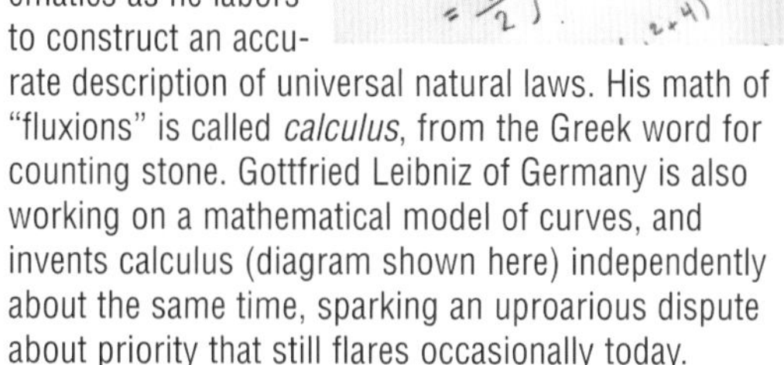

1666 Isaac Newton of England devises a new system of mathematics as he labors to construct an accurate description of universal natural laws. His math of "fluxions" is called *calculus*, from the Greek word for counting stone. Gottfried Leibniz of Germany is also working on a mathematical model of curves, and invents calculus (diagram shown here) independently about the same time, sparking an uproarious dispute about priority that still flares occasionally today.

1679 Denis Papin, student and colleague of Christiaan Huygens, investigates novel theories about raising the boiling point of liquid by heating it under pressure. Invents a safety valve, making the pressure cooker possible (Papin calls it the "steam digester"). Real import of achievement: basis of the Age of Steam and the steam engine.

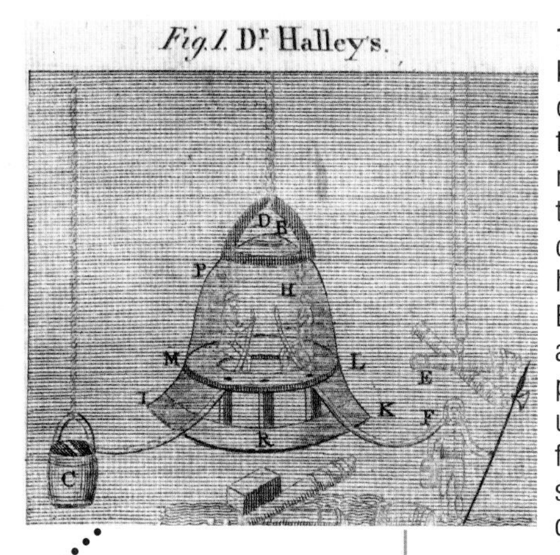

1690 Edmond Halley, best known for predicting the return of a reappearing comet, invents the diving bell, with a hoisting mechanism like those still used to raise and lower submersible craft. Alexander the Great had lowered himself into the Bosporus in a glass jar to look at fish, but Halley's invention permits long, repeated stays underwater. Proves impractical for other purposes, such as salvage, construction, or dredging.

c. 1685
J. C. Denner of Germany expands the capacity of the medieval chalumeau, a wind instrument, by lengthening it to 3 octaves, and invents the first modern reed instrument: the clarinet.

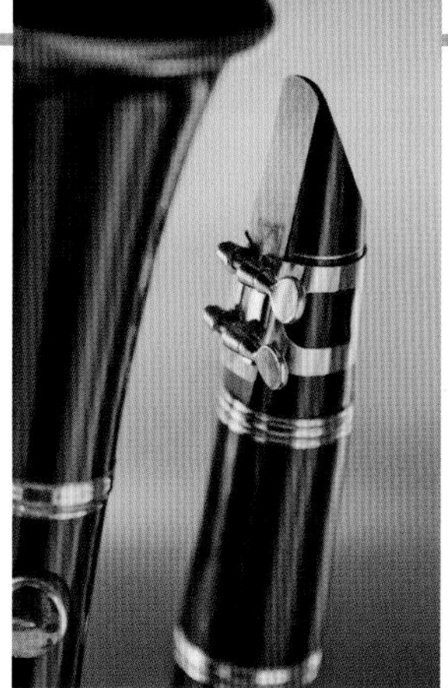

Birth of the Orchestra

Other instruments debut in 18th century: the mandolin (1744) and 6-string guitar (1779), the earliest surviving examples of which are made by Gaetano Vinaccia of Italy. Both evolve from the lute, which traces back millennia to single-stringed African instruments. As Classical era arrives, Stradivarius manufactures his finest violins, violas, etc., 1700–1720.

1700

1709
Bartolomeo Cristofori of Italy invents the piano, or *gravecembalo col piano e forte*—"harpsichord with soft and loud." Slow to become popular in Italy, by the middle of the 18th century, the piano gains acceptance in Germany and elsewhere.

1701 Gentleman farmer Jethro Tull addresses one of farming's great problems: waste of seed and effort due to the "broad-casting" method of sowing. His seed drill punches holes in soil and drops seed in from hopper; planting in rows instead of randomly makes hoeing for weed control far easier. But technique takes decades to catch on. Unknown to Europeans, less advanced versions of the drill have been used in Asia for centuries.

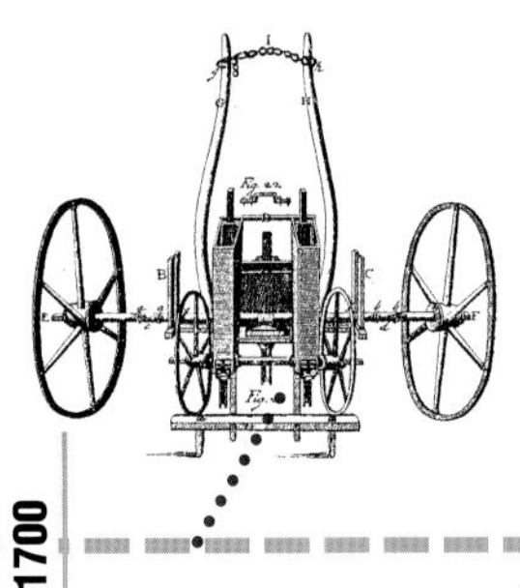

1709 Abraham Darby of England succeeds in smelting iron using coke, a by-product of baking off gasses and tar from coal. The Darby method allows the cheap product of high-strength iron for casting; his grandson uses this method to build the first cast-iron bridge, seen here, in 1779. Cheap production of high-strength iron major step toward Industrial Revolution.

Collective Invention

Steam engine is perhaps first, best example of how inventions, their nature, and the principles they employ become more complex. Simply put, no one is the sole inventor of the steam engine; it is invented in stages of innovation over decades. Papin arguably the first to make a piston move with steam; Thomas Savery's engine does the first real work (water pumping) with steam; Thomas Newcomen makes the Savery engine much more efficient and his plans are widely adopted; and James Watt's novel engine components make engines most efficient, powerful, and useable in factories, boats, and trains. This evolution in increments from laboratory to workability is the new norm in progress.

1700

1707 Denis Papin uses his knowledge of steam technology to affix an engine (a very weak one) to a paddle wheel, inventing the steamboat. It meets an immediate, unfortunate fate: boatmen threatened by specter of job displacement destroy his craft. The Marquis Claude de Jouffroy cannot raise enough financial support to market his viable steamboat in 1783; American John Fitch patents a workable but too-expensive steamboat in 1787. World embraces technology with Americans Robert Fulton, whose *Clermont* is seen here, and Robert Livingston in 19th century.

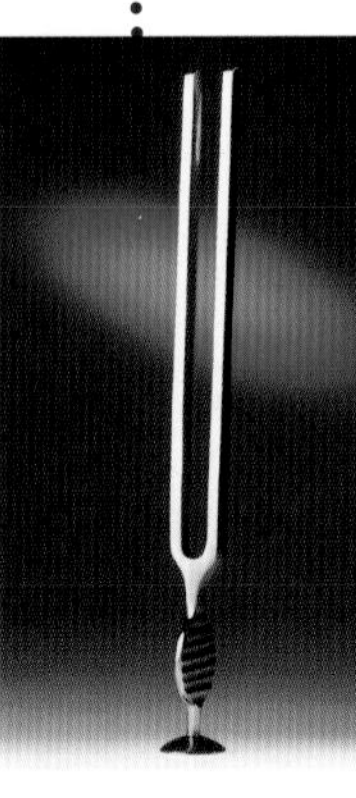

1711 Tuning fork invented by John Shore, sergeant trumpeter to the court of England, who's looking for methods to produce a reliable tone for instrument tuning. Shore's simple invention supplants the pitch pipe, a curved piece of metal with straight handle that when struck produces a reliable A note by which musicians can calibrate their different instruments. He reputedly tells an audience before using his new device publicly for the first time: "I do not have about me a pitch-pipe, but I have what will do as well to tune by, a pitch fork."

c. 1730 Sextant added to set of tools, profoundly improving navigation. Invented independently in the same year by John Hadley of Britain and Thomas Godfrey of the American colonies. Angled mirror is moved on an arm to bring image of a celestial object in line with the horizon. At that moment, height of the object in the sky is read in degrees on curved scale of sextant. Sextants show exact latitude and local time critical to determining longitude. Hadley's device shows 45 degrees; Godfrey's sextant can show a sixth of circle or 60 degrees.

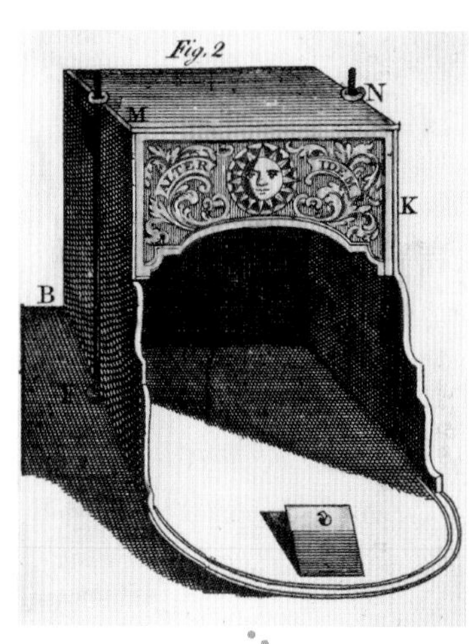

1742 Benjamin Franklin creates the wood-burning Franklin stove, which fits into a fireplace and creates more heat while reducing fumes. The stove is an immediate hit, and an improved coal-burning version in 1770 adds to the American genius's immense popularity and stature.

1712 Thomas Newcomen improves ineffective steam pump built by Thomas Savery. Cylinder of alternately boiling and condensing water moves piston back and forth—first true steam engine. James Watt of Scotland makes great improvements starting in 1765 and continuing for many years, and is often called the inventor.

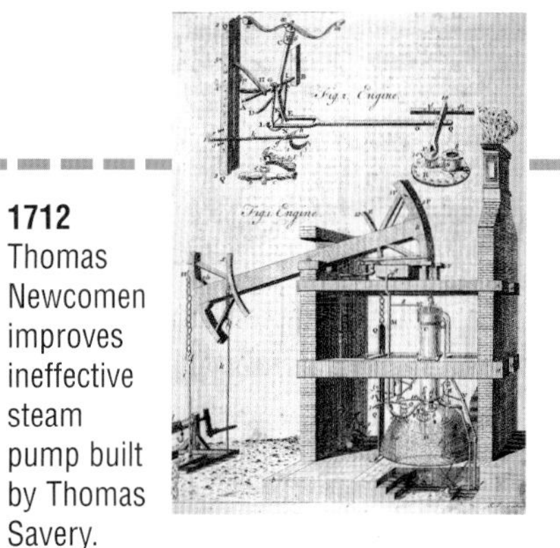

1733 Flying shuttle, which greatly extends the width and color variety that can be woven on a hand loom, patented by John Kay. The shuttle, used by hand weavers since antiquity, carries yarn or thread back and forth crosswise, becomes cumbersome for broadloom weaving. Kay's innovation is to rig the shuttle to cords, allowing a single stationary weaver to work much more efficiently. Paves way for industrialized textile creation. Kay loses everything trying to protect his patent against profit-hungry manufacturers.

Generous Genius

"As we enjoy great advantages from the inventions of others, we should be glad of an opportunity to serve others by any invention of ours; and this we should do freely and generously." So wrote Benjamin Franklin, who never sought a patent for any of his inventions, and declined the offers of others to do so on his behalf. He did support the creation of the U.S. patent system, though.

1746 E. G. von Kleist of Germany and physicist Pieter von Musschenbroek of Leyden, Holland, independently invent an effective way to store up and discharge electrical energy—the first capacitor. The "Leyden jar" consists of a conducting wire in a jar of water, held in place by an insulating cap. Later innovations lead to the first battery.

1750 German, Swiss gun makers in Pennsylvania perfect method of grooving, or "rifling," spirals on the inside of gun barrels. These weapons shoot straighter, farther, with less weight needed in the barrel. Called "Kentucky rifle," probably for Daniel Boone, the Kentucky frontiersman (originally from Pennsylvania) who extolled their superiority.

1746 After seeing a demonstration of static electricity in London, Benjamin Franklin tries to attract lightning by mounting a long iron spike on Christ Church in Philadelphia. Though Franklin's goal is to capture an electric charge, the safety advantage of "lightning rod" is immediately obvious.

1762 British banker Lawrence Childs devises a way to decrease fraud as checks become an increasingly common method of transferring funds. He prints a stack of checks, numbers them sequentially, and binds them, inventing the checkbook and modern checking accounts.

1763 John Harrison of England claims prize set 46 years earlier by the British Board of Longitude for the first person to invent or discover a precise, consistent method of tracking horizontal location—longitude—on the earth's surface. Harrison, ignoring others' attempts to use heavenly bodies for tracking, constructs a large, highly accurate watch after 25 years of experimenting and 3 scrapped designs. Carried aboard ship and faithfully keeping London time exquisitely, chronometer allows mariners to know their longitude by comparing local and London time; solution to perhaps the most important technological problem of the era, one that had cost many lives and much money.

Brilliant Timing

Briton John Harrison, a Yorkshire joiner (carpenter) with limited education but a brilliant mind, builds his first clock (entirely of wood) at age 20. He and his brother achieve unparalleled levels of timekeeping accuracy when he devises the chronometer, method for determining precise longitude. Political enemies thwart his rightful claims to the Longitude Prize, requiring King George III's intervention to get him even the partial payout he is eventually awarded. Along with sextant, chronometer permanently solves the problem of uncertain navigation; the combination is not significantly improved upon for 2 centuries.

1764 James Hargreaves 1st to meet challenges of the Royal Society of Arts; invents a weaving machine that turns 6 spinning threads into yarn, with only 1 person powering the machine. The "spinning jenny" outrages hand weavers, augers Industrial Revolution.

1766 British engraver and mapmaker John Spilsbury cuts one of his world maps into pieces and mounts each piece on wood. The jigsaw puzzle becomes educational tool for teachers and new pastime around the world.

1767 Joseph Priestley, one of the discoverers of oxygen, invents a way to dissolve carbon dioxide in water under pressure. Soda water is born.

1769 Richard Arkwright patents a machine that spins cotton fibers into yarn with an external power source, first a horse on a treadmill, then water power. The "water frame," with 1,000 spindles per machine, opens a new era in industry.

Happy Ending

Cugnot's life story puts happy twist on the all-too-common tale of a brilliant inventor who dies in poverty. After winning acclaim for his tractor and other inventions, he is banished to Belgium during the French Revolution and spends years bereft and in obscurity. But in old age, Napoleon grants him a pension; he dies in Paris in 1804.

Fabric of Change

Along with flying shuttle and spinning jenny and, later, the spinning mule, technology transforms the textiles industry and sparks profound social changes as the factory becomes the center of life for many workers and communities. Corporations form to raise funds for costly equipment, with investors profiting. Modern capitalism begins in textile factories.

1769 Nicolas-Joseph Cugnot, an engineer in the French army, makes improvements on the steam engine and deploys them in a steam-powered tractor: world's 1st self-propelled vehicle and arguably the 1st automobile. Literally hits a brick wall in 1770 in perhaps first auto accident. Cugnot's tractor likely the 1st conversion of straight engine-powered motion into rotary motion (via the wheels).

1772 Henry Clay of Birmingham mixes varnish into a cottony paper and hearkens birth of modern papier-mâché, though it was originally invented by the Chinese in the 3rd century BC. Clay's new material sweeps Europe and becomes all the fashion for masks, art, decorations, and even furniture.

1774 Georges-Louis Lesage takes first step toward electronic communications by inventing original telegraph, a series of 26 wires with letter keys on 1 end and needle indicators at the other. Range of transmission: adjoining rooms in Lesage's home, yet a significant advance nonetheless. Design proposed by Francisco Salvá requires 26 servants, each assigned a letter of the alphabet, to get shocks at the receiving end of a wire and to shout out their letter.

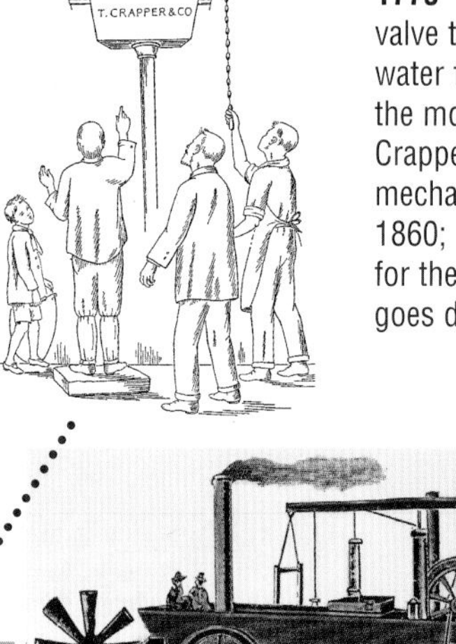

1775 Alexander Cummings patents a valve that reliably holds and then releases water from a flush tank: key component in the modern toilet. British plumber Thomas Crapper devises an automatic shutoff mechanism for the Cummings valve in 1860; he winds up stealing all the glory for the whole invention, and his name goes down in history.

Electricity

Along with the harnessing of steam power, exploration of electricity was the technological work of the 1700s that most meaningfully set the stage for the modern world. Invention of the Leyden jar and experiments such as in France, where 1,000 monks held a wire and all jumped as a charge was passed through them, hinting at the potential power waiting to be exploited.

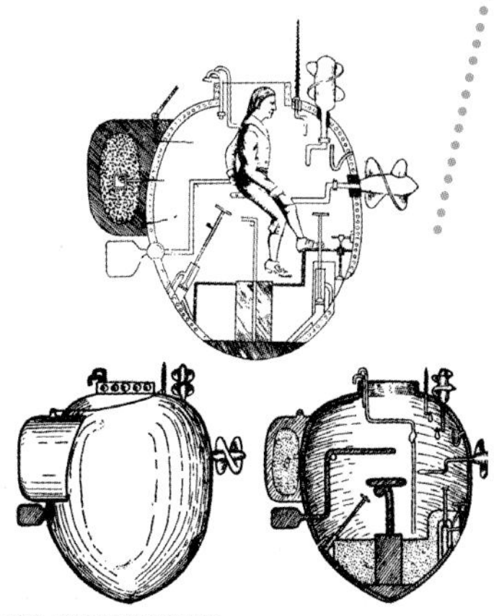

1775 Seven and half feet high and 6 feet wide, the 1st submarine used in combat, the wooden *Turtle*, is unveiled by inventor David Bushnell of Connecticut. Operator enters through top hatch, uses hand cranks and rudder to navigate, and resurfaces to pump in new air. In 1776, *Turtle* performs superbly in water on 1st military assignment, but is unable to drill hole in British ship's hull to attach keg of powder. Mission fails.

1777 American mechanical genius Oliver Evans comes out with the 1st in a lifetime of useful inventions: a machine for making teeth for carding paddles, used to straighten the fibers of cotton or wool. Evans's later achievements include the grain elevator and an automatic flour mill; the high-pressure steam engine crucial to industrial development; and the Orukter Amphibolos ("Amphibious Digger"), seen here, a steam-run river-dredging boat with wheels that performs poorly but is nonetheless the world's first amphibious vehicle.

1777 First patent for circular saw granted to Samuel Miller of Southampton, England. Continuous cutting action replaces back-and-forth motion of existing sawmill blades. Productivity and deforestation both increase.

One-flight Wonder

In the 18th century, Montgolfier is the first name in air travel because the brothers Joseph and Étienne are first to transport a person successfully through the air. But they are not the first people to think of and construct a hot-air balloon. Largely forgotten: In 1709 Brazilian priest Bartolomeu de Gusmão demonstrates to the Portuguese court his own small, working hot-air balloon. He is rewarded with a professorship. Yet he and his invention are immediately forgotten, and the Montgolfiers are accorded full credit for the idea; in French, *montgolfiere* means hot-air balloon.

1784 Weary of switching between the glasses he needs for reading and the ones he uses for long-distance vision, Benjamin Franklin cuts both pairs in half. Mounts lower halves of reading glasses together in same frame with upper halves of far-viewing spectacles. Invention of bifocals.

1783 Joseph Montgolfier, son of a wealthy paper manufacturer, watches as a paper bag rises over a fire. He and brother Étienne construct paper-lined cloth bag that carries a flame, leading to hot-air balloon. In 1783 their balloon rises 1,000 feet and travels a mile. First untethered human air flight occurs when François Laurent and Pilâtre de Rozier stay aloft for more than half an hour. In 1785, Pilâtre and companion Pierre Romain become first aviation fatalities.

1784 Joseph Bramah invents the tumbler system for locks. Though lock making is known in antiquity, this is the 1st design that can't be easily broken. Places a sign in his shop reading, "The artist who can make an instrument that will pick or open this lock shall receive 200 guineas the moment it is produced." The reward, $50,000 (in 2007 U.S. dollars), stands uncollected for 67 years.

1784 One of Europe's leading intellects, Aimé Argand of Switzerland, takes on the problem of the oil lamp, largely unimproved for centuries. He and his team reinvent the wick, make it cylindrical for more complete combustion, and place it in a circular holder that draws up air through rings of holes feeding both inside and outside of wick. Separate oil reservoir delivers fuel. The circular-burner design Argand pioneers still seen in design of modern gas stoves. Argand lamp an immediate success; markedly enhances the brightness level of indoor spaces worldwide.

1784 Scattershot's usefulness as only a short-range ammunition prompts Lieutenant Henry Shrapnel of the British Royal Artillery to devise a shell that has a delay fuse, designed to burst over troops at long range. First used in combat in 1804 and extremely effective. Shrapnel promoted, rewarded with lifetime pensions; his name still used for scattered, lethal debris (technically, only the pellets are shrapnel).

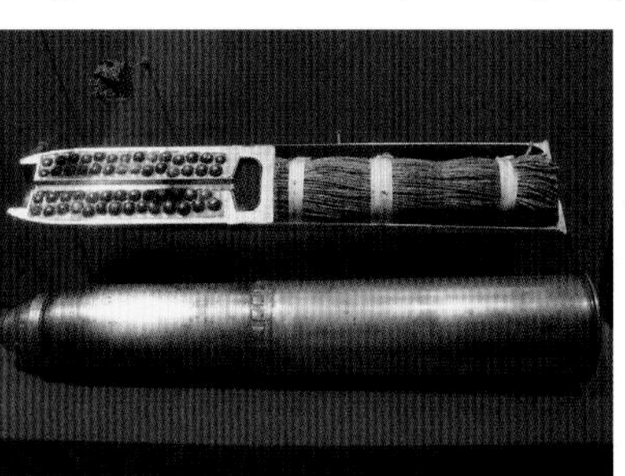

1785–1790 Arkwright's genius with the "water frame" (see year 1769) has created a strange problem: cotton yarn can now be spun so quickly and efficiently, hand looms can't keep up. Edmund Cartwright automates the motions of the weaver; his power loom is the basis for the full mechanization of the fabric industry.

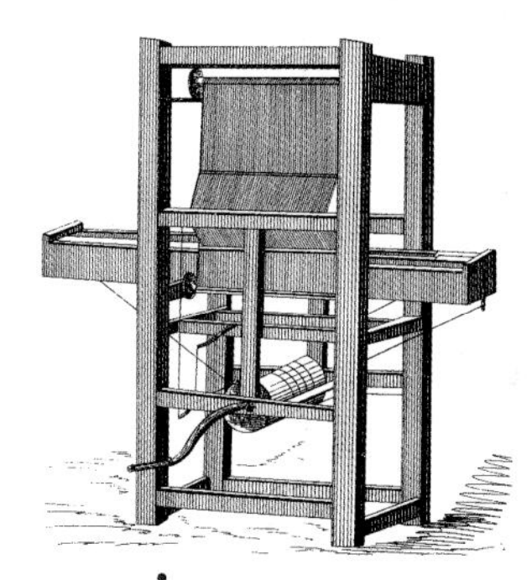

1785 Jean-Pierre Minckelers, university professor, invents gas lighting when he develops a process to distill coal and capture the emissions. His process, resulting in first street lighting, serves as basis of lighting systems that enhance public safety. As a practical matter, gas lighting is a collaborative effort of numerous inventors; first noteworthy systems of lamps installed in England by William Murdock, in cotton mills and private buildings; 5 years later, Philippe le Bon demonstrates his system for the public in Paris. In 1807, first municipal system of piped gas installed in Britain's Pall Mall.

1785 Oliver Evans studies the operations involved in milling flour, and patents an astonishing assemblage of 5 separate machines to take the place of manual processes. Proceeds to write *The Young Mill-wright and Millers' Guide*, the definitive and highly influential book on powering the mill industry.

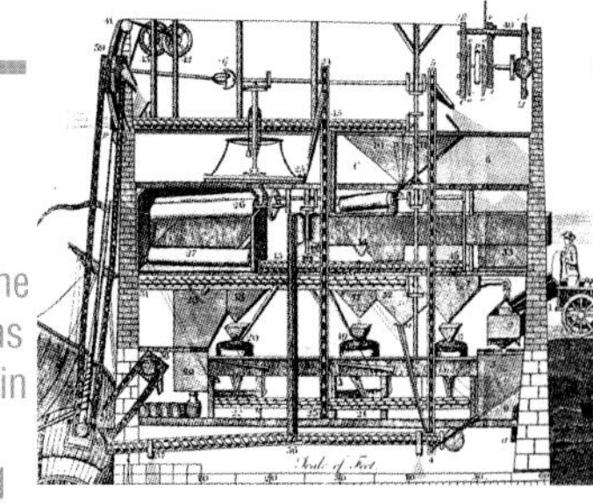

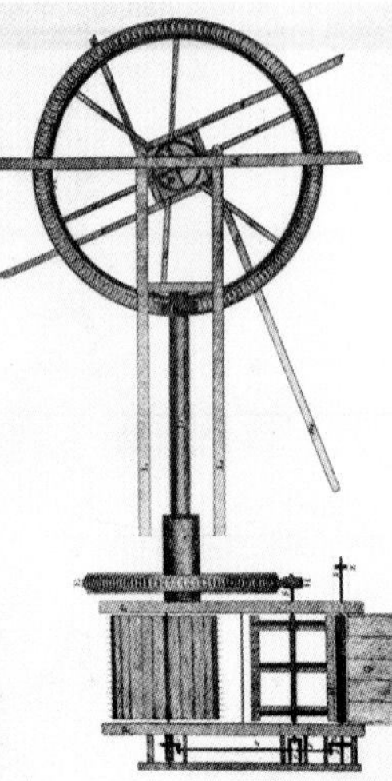

1786 Andrew Meikle patents automatic thresher, which separates grain from its stalk and chaff via a series of vertical blades housed in a drum. Widespread adoption of the machine leads to riots, quickly suppressed, among farm laborers in England.

1787 Claude Louis Berthollet, one of France's leading chemists, invents the use of chlorine to bleach fabric white; process greatly advanced by work of Scotland's Charles Tennant, who in 1799 invents a powder that makes possible the widespread use of bleach for paper and cloth.

1788 Article One of U.S. Constitution grants Congress the right "to promote the progress of science and useful arts by securing for limited times to authors and inventors the exclusive right to their respective writings and discoveries." Washington asks Congress to pass a patent law in 1790, and the U.S. Patent Office opens that year.

1789 Englishman Samuel Slater, memorizing the workings of Richard Arkwright's textile machinery, moves to America and adapts the technology to suit the supplies, materials, and parts available in the U.S. Located in Pawtucket, Rhode Island, his is the 1st factory of the Industrial Revolution in America.

Fabric of the Industrial Revolution

Pioneers in the automation of the textile industry solve different problems at different times, but all have one thing in common: the negative and often violent reaction of workers in the existing industry. Cartwright's first truly effective weaving factory opens in 1790; it burns down suspiciously the next year. Violence culminates in early 1800s with Luddites, who smash machines across Greater London to protest worsening economic status; "Luddite" still means one who resists new technology.

1790 Jacob Perkins of Massachusetts completes the most significant invention of his distinguished career (49 patents): a machine that cuts a nail and makes its head in 1 operation. Major improvement in process. But like so many innovators, he cannot protect his patent or profits.

1790

1791 French Nicholas Dubois de Chemant causes a sensation with his new method for fitting false teeth. For millennia, they've been wood or bone, but Chemant uses porcelain mineral paste to create long-lasting and attractive bridgework. France's General Comte de Martagne writes the following verse in honor of Chemant:

When time has stripped our armoury bare
Dubois steps in with subtle heed
New grinders and new cutters gives;
With his we laugh, with his we feed.
Long live Chemant, friend in need.

1792 As the French Revolution creates a booming execution trade, Dr. Joseph-Ignace Guillotin decides the old, brutal methods of putting prisoners to death are inhumane and unfit for the new republic. He improves existing devices that use a weighted blade launched down a frame to behead the condemned. Guillotine becomes symbol of the dark excesses of the Revolution.

1792 Charles Chappe introduces semaphore (optical telegraph), system of towers with moveable arms controlled by ropes and pulleys. Placed on hilltops about 6 miles apart, the towers relay information 90 times faster than a horse can carry it; within a decade, semaphore towers blanket France. The technology endures worldwide until advent of electric telegraph in the 1840s. Semaphores are visual signals that are still in use today.

Age of Enlightenment

Late 18th century sees full fruition of the philosophy Francis Bacon sets out in the early 1600s. He advocates using reason instead of blind acceptance of dogma through scientific methods of testing, observing, and reporting. The approach spreads to all facets of social endeavor (politics, philosophy, the arts) in the period 1760–1800, also known as the Age of Enlightenment. By the turn of the century, the belief in a rational, and therefore discoverable, underpinning for all existence is summed up by theologian/biologist William Paley, who calls God the "great watchmaker."

Machine Tools

Work on electricity and steam indicates how the power for large-scale industry could be generated, but a different kind of inventive genius provides the other element of the coming Industrial Age: precision. Machine parts that spin quickly, cut cleanly, and perform intricate operations over and over without breaking down have to be superbly constructed; this is mostly an art of fine and repeatable metal cutting. Maudslay is a pioneer in this new craft of machine tools. An apprentice to Joseph Bramah, who made the first unpickable lock, Maudslay makes his mark by improving a metal lathe in Bramah's shop in 1792, and his own workshop becomes the cradle of the whole machine-tool industry, without which the modern world could not have been manufactured.

1793 Hannah Slater, wife of the mechanization pioneer Samuel Slater, develops a process to spin very fine cotton thread—a less expensive substitute for linen thread. Becomes first female patent holder in America.

1794 British ironsmith Philip Vaughn invents the ball bearing as we know it—metal balls racing around a circular track. Makes a panoply of modern rotating components possible, from skateboard wheels to gun turrets.

1793 The widow of Revolutionary War hero Nathaniel Greene notices how handy her houseguest Eli Whitney is, and complains about an urgent problem. The only existing device to remove seeds from cotton won't work with long-fiber American cotton. Whitney's invention—a single mechanism run by a hand crank—combines teeth for seed removal and a brush to extract cotton fibers. The cotton gin is born, dramatically increasing the demand for cotton and for slave labor in the South.

1794 Henry Maudslay patents, designs, and builds the first really precise clamping system for cutting metal with standardized threads. For practical purposes, this is the invention of the standardized nut and bolt.

c. 1795 Officers of the French army press for every rifle part to be identical for quick replacement from another weapon during combat. "Cannibalization" system first step toward interchangeable parts. Past practice, such as for building wheel-lock rifles seen here, required custom parts for each weapon.

1797 After investing $30,000 (almost $3.5 million in 2007 U.S. dollars), Charles Newbold of New Jersey patents method for casting single-piece iron plow; invention greatly eases groundbreaking but rejected by the market because farmers fear poisoning their soil. Newbold loses investment.

Whitney's Show

Eli Whitney demonstrated value and viability of interchangeable machine parts in 1798, even though his idea wasn't fully realized until about 1830. In an appearance before the U.S. Congress, he disassembled rifles of his own manufacture, mixed their parts, then reassembled them from the same parts randomly chosen. "Remove and replace" permitted gigantic increase in productivity, from small farms to enormous factories, and was essential portal to Industrial Age.

1796 Edward Jenner, an English surgeon, tries an experiment that rids the world of its worst health menace, smallpox, and confirms the value of inoculation. Jenner notices that milkmaids, who almost never get smallpox, get cowpox. Jenner injects 1, and then about 20 other children, with cowpox serum and, later, live smallpox virus. None of the children get smallpox. The disease, which kills an estimated 20 percent of the European population during some outbreaks, is wiped off the planet over next 2 centuries. Variolation—deliberately infecting a healthy person to induce immunity—had been practiced for centuries, but Jenner invents a medically valid form of the technique that transforms public health.

1797 Originally proposed by Leonardo da Vinci in late 1400s, parachute is refined by André-Jacques Garnerin, who jumps from a balloon at 3,000 feet in a basket fitted with a silk umbrella-like canopy. Garnerin gets motion sickness on descent, but the device works. Vents for stability (1st demonstrated by Garnerin in a subsequent jump), and folding models come later.

From Interchangeable Parts to the Airplane: 1798–1903

Observers who had stood on a street corner near the center of New York, Paris, or Tokyo in 1800 would have encountered a reinvented world, had they returned again in 1900. In 1800, people traveled at the speed of horses, read by candlelight, waited agonizing weeks to hear from loved ones far away. They sewed by hand and made their own soap. A toothache could be a death sentence. But an eyeblink later in historical terms, this had all changed forever and for the better. Precisely tooled machinery came together with steam power early in the 1800s, yielding the steamship and locomotive. By century's end, steam, electromagnets, and turbines were combined to light up the world with the magic of electricity. Better instruments and research methods were creating profound breakthroughs in medicine: from anesthesia to antiseptics to aspirin. All these changes took place an invention at a time, but at an unbelievably fast rate.

The previous few centuries had seen tremendous achievements in science, but much of that progress was academic. Knowing that the earth revolved around the sun didn't heat anyone's home or save their sick baby. But with the start of the Industrial Revolution, the products of science crossed over into the realm of the practical and everyday. A cavalcade of useful, helpful, and enjoyable devices and processes made their way from the inventor's workshop and factory floor into the villages and households of the world. Celluloid, brilliant new dyes, chewing gum, movies, bicycles: Each new product of the age seemed more wonderful than the last, and in contrast to the microscopes and chronometers of the preceding era, these inventions were meant for everyone. There were the usual setbacks, backlashes, and bitter disputes, but the century was a kaleidoscope of amazing and delightful inventions—including the kaleidoscope itself!

It all started with steam. The windmill and waterwheel had already spurred a science of cranks, pistons, and gears, and these had been studied and improved over centuries. As the physicists of the Scientific Revolution became familiar with the properties of water vapor under pressure, they soon applied it to all the moving parts of the earlier technology, and started solving the problems of the new one. When engines that exploited the power of boiling water were made reasonably efficient and reliable, they were put to work accomplishing miracles: pumping, grinding, spinning, and turning at rates totally undreamed of in human experience. The distance between Chicago and New York remained the same (790 miles), but in travel time it plummeted from three weeks to two days.

Nowhere was steam's world-shifting power better demonstrated than in the generation of electricity. Static electricity had already been generated at the end of the seventeenth century, and throughout the eighteenth century scientists like Benjamin Franklin became better and better acquainted with its nature. Right at the turn of the nineteenth century, the first current was produced chemically—that is, with batteries. A few decades later, the British genius Michael Faraday demonstrated that electricity is—literally—in some ways the flip side of magnetism. An electrical current can cause a nearby magnet to spin, and this became the basis of the electric motor. Soon thereafter, Faraday showed that a conductive metal in a charging magnetic field will produce a current. That led to the rapid development of generators, and by 1895 huge dynamos were anchoring the world's first power grids.

Well before that, battery-produced electricity had totally transformed communications. Experimenters worldwide, working with the transmission of current over wires, figured out how to send pulses of current reliably, yielding the first instant long-distance communications technology: the telegraph. In the United States, the patent was issued to Samuel Morse, but one historical study found that at least sixty people could legitimately claim to be the inventor. Now, messages could travel at the speed of light. Later in the century, the electrical engineering underlying Morse's system evolved into the telephone.

Whether or not Morse deserved so much of the credit for the device he patented, he certainly got it. Morse was just one of a new breed of hero-inventor that arose in the century. Benz in Germany; Pasteur in France; Edison, Bell, and the Wright brothers in America: They were all lionized, almost deified, by a popular press more available to more people than ever before. And led by Edison, this generation of giants also invented the deliberate research-and-development process itself.

While the auto, phone, and medical advances had the biggest impact on the personal level, changing society one person at a time, another enormous theme of the age was enormity itself—mass production and mass consumption, along with increases in scale of every sort. Railroads spanned Europe and America. Buildings rose hundreds of feet above street level, made possible by water pumps, steam heating, and the safety elevator. Factories employed thousands instead of dozens. And it was in this sphere, the realm of mass manufacturing, that the century really left its mark.

Interchangeable parts meant that finished products could be repaired and replaced easily and quickly, but more importantly, the machines to make those products and parts also could. That meant products from rivets to ribbons could be turned out cheaply, by the millions, on giant production lines.

All this hugeness and abundance came with a price. Whole battalions of soldiers now perished in minutes as horrible new weapons made killing more efficient. And some saw a subtler tragedy looming in the new ways as the skies filled with smoke, children took their places on dangerous and soul-deadening factory floors, and small-scale artisans vanished beneath a torrent of big companies. From the beginning of this period, bright visions of utopia collided with the reality of what William Blake called "the dark Satanic mills."

But if you had asked the street-corner observers whether they wanted to go back to 1800, they very likely would have said no. In the real world, despite the "Luddite" attempts to stop the advance of new machines, and despite the lamentations of the *New York Times* reporter who detailed how cheap tin horns "of the recently-introduced type" drowned out the bells of Trinity Church at the first midnight of the new century, any question was pointless. There was no going back.

1798 Austrian playwright Aloys Senefelder wants to publish his own work, but it's too expensive, and he's experimenting with new printing methods. One day, using a grease pencil, he scribbles a laundry list on the closest item at hand: a piece of limestone in his studio. That list sets in motion a chain of experiments that leads to lithography, still one of the world's most valuable printing methods. It exploits the antipathy of oil and water to put ink on a page in an inexpensive, attractive way.

1801 Joseph-Marie Jacquard adopts a technology invented earlier by Jacques de Vaucanson and builds a loom that can repeat complicated weaving patterns infinitely. It receives operating instructions from a series of holes punched into heavy paper—a programmable machine. Workers destroy several Jacquard looms, but the machine is a huge hit with industry.

1800

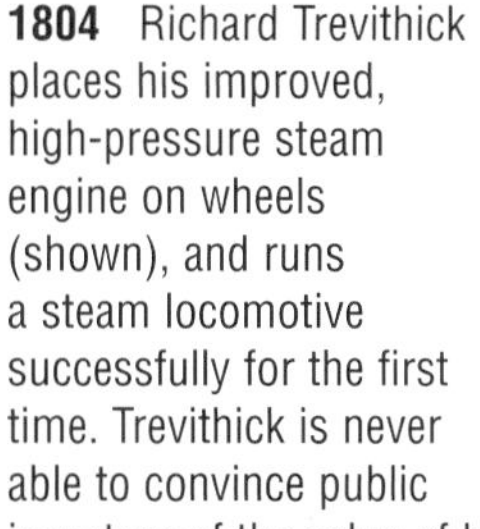

1801 Alessandro Volta disagrees with his friend Luigi Galvani, who's convinced he's found "animal electricity" because a frog's legs jump when a charge is passed through them. Volta begins experimenting with chemical reactions to see if he can create current; builds stacks of alternating copper and zinc strips immersed in salt water. These do create a charge, and the electric battery is born. Volta soon constructs "voltaic piles" (seen here) of copper and zinc interspersed with salt-treated cardboard; these are the first dry batteries. Start of a 2-century process of improvement that is still a major research focus as companies search for better batteries for equipment from hearing aids to spaceships.

1804 Richard Trevithick places his improved, high-pressure steam engine on wheels (shown), and runs a steam locomotive successfully for the first time. Trevithick is never able to convince public investors of the value of his device; low-strength rails hinder acceptance. George Stephenson of England makes trains really useable; introduces the flanged wheel to keep locomotives "on track"; and rails get stronger. At the 1829 "Rainhill trials" in England, the Stephenson machine proves its merit, and the Age of Rail is really born.

1807 Robert Fulton sails his *Clermont* steamship upstream from Manhattan to Albany in 32 hours, causing a worldwide sensation and proving steamships work. For the 1st time in history, sea and river travel do not depend on wind and current, and boat designers are freed from planning all their work around capturing wind. After many years of working on submarines, Fulton listens to his advisers and switches to steamboats, building on a century of work and culminating in his 1807 inaugural trip.

Mass Production

In 1802 Henry Maudslay, genius of machine tools that hold and cut with tolerances of 1/10,000 of an inch, set up a production line to meet the demand of the British navy for 130,000 ship pulleys ASAP. He networked 45 machines in sequence, each one performing just one task (such as cutting a groove). Multiplied the efficiency of each worker tenfold, but also meant workers were no longer crafts-people, a downside that worsened as mass production techniques evolved toward the assembly line. Maudslay, a mechanical genius in the true sense, never came up with a startlingly new idea, but he brought to life the ideas of others even when they had long proven "impossible." Eli Whitney converted Maudslay's work into "the American system" of mass production, and American engineering know-how eventually superseded England's; throughout the 19th century the new nation became the industrial and economic powerhouse of the world.

1810 Peter Durand creates the tin can, and the canning industry. In 1795, after Napoleon offered a 12,000-franc reward to the person who could devise a method of preserving food for use on the battlefield, Nicolas Appert boils glass containers filled with food and seals them—invention of sterilization. Though Appert cannot explain why, his method stops spoilage. Durand seeks an unbreakable form of the product; coats iron canisters with tin, sells his patent to Bryan Donkin and John Hall, who open the first cannery in 1812.

1812 German organ maker Dietrich Nikolaus Winkel mounts a pendulum upright so the weight is at the top, with a driving weight at the bottom to keep it swinging. The metronome keeps a very accurate beat, so musicians can play music more closely to how it is written. This is important to composers such as Beethoven, who incorporates metronome marks into his work almost as soon as the device is operable. One of the metronomes by J. Maelzel, who patented Winkel's device, appears here along with Beethoven's manuscripts.

1816 Sir David Brewster of Scotland, a trailblazer in the study of light who made important discoveries about polarization and diffraction, arranges 2 mirrors in a tube and seals bits of colored glass at 1 end. Brewster patents the kaleidoscope in 1817, but can't hold on to its profits because a flaw in the patent allows other people to take the idea.

1816 After a series of horrible methane explosions in England, nonexplosive lanterns are invented by Humphry Davy and George Stephenson. Stephenson's work is tested October 1815, Davy's some months later. Both lamps work by diffusing "firedamp" (methane gas) so that the lantern won't explode. Both men refuse to profit monetarily from their lamps, but they fight over credit for the rest of their lives. Ironically, the new lamps lead to deeper, riskier mine exploration—and more disasters. (Shown, modern lamp.)

Brilliant and Bitter

The Davy-Stephenson feud was one of the most fascinating in the long history of disputes between great inventors. Both were already famous and respected: Stephenson for his work on railroads and locomotives, Davy for discovering laughing gas for anesthesia, for inventing the arc lamp, and for his brilliant work isolating elements such as potassium, sodium, and chlorine. Both were motivated by altruism and compassion in their attempts to halt the scourge of mine disasters, yet strikingly petty in trading accusations and denials against each other. Some historians believe the dispute is partly fueled by class issues: Davy was a knight and celebrity, Stephenson came from a poor mining-country family. Both died still bitter about their battle.

1816

1816 René Laënnec, one of France's most respected doctors, avoids the traditional method of examining his patient, an obese young woman, which calls for laying his head on her chest and abdomen to have a listen. Out of distaste or modesty, he rolls a sheaf of paper into a cone and holds the wide end against her body, the small end into his ear. Laënnec is astonished at how well he can hear—and so is the medical community, which still uses the stethoscope as its first diagnostic tool.

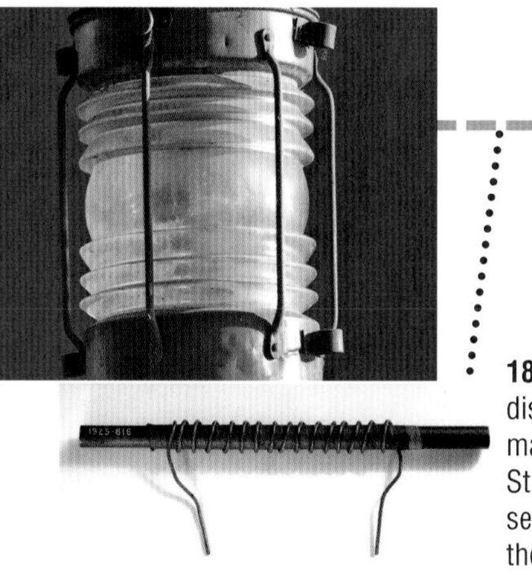

1820 Spurred by Hans Christian Oersted's discovery that electric currents seem to attract magnetic compasses, British physicist William Sturgeon wraps a wire around an iron bar and sends a charge through the wire. As he suspects, the bar turns into a magnet for as long as current flows through the wire. Sturgeon confirms electromagnetism and invents the solenoid, which proves vastly useful in switches and machines over the centuries. By 1831, American Joseph Henry lifts more than a ton of iron with his electromagnets, and moves a metal rod from a mile away, hinting at the possibilities for relaying electrical signals across long distances.

1826 Attempting to find uses for the waste products of natural gas production, chemist Charles Macintosh mixes rubber and naphtha, a residue from petroleum distillation, which makes fabrics waterproof. Slathering his new concoction around an overcoat, inside and out, he invents the "mackintosh," a gigantic success.

1827 In 1821, French artillery officer Charles Barbier demonstrates his "night writing," or system of embossed dots, to students at the National Institute for the Young Blind in Paris. One of the students, Louis Braille, recognizes its potential at once, and after 6 years has simplified and improved upon Barbier's writing system, making it ideal for the visually impaired—for writing, not just reading, which is the real value of the Braille method. In the past, blind people were all but unable to write, because the methods were so laborious to learn and use. The first Braille book, a history of France, comes out shortly after Braille perfects his alphabet and writing stylus.

1826 Joseph Nicéphore Niepce finds a mixture of chemicals (bitumen, oil of lavender) that will capture a scene focused through a lens onto a pewter slab and preserve it forever. Niepce follows the well-worn, frustrating path of many an inventor when others fail to recognize and reward what he's accomplished. Searching for support, he forms a partnership with Louis Daguerre, who makes considerable improvements in the chemistry and processing of photographic images. Niepce dies in 1833. Daguerre, who thinks enough of his own work that he bestows his name on the new marvel, the daguerreotype, unveils his successful process in 1839, to worldwide acclaim.

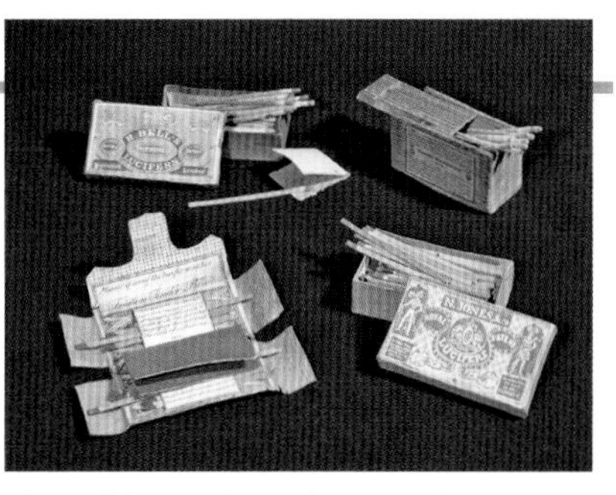

1827 Englishman John Walker invents the match by combining antimony sulfide and potassium on a wooden splinter, and a striking surface of coarse glass on paper. This is first portable, easily ignited flame. By 1911, after many people are poisoned by phosphorous in their matches, the Diamond Match Company bows to pressure from President Taft and surrenders its exclusive patent rights to a nontoxic formula, for the good of the public.

1827 Josef Ressel of Austria puts his flash of insight to use. He realizes a turning screw produces force in the opposite direction of its turning, so why not use it to drive water backward or a boat forward? He fits a screw propeller on the keel, and it works. In 1834, Britain's Francis Pettit Smith experiments with same idea, but his propeller breaks and loses a half turn of the screw—and the boat speeds up. True birth of the screw propeller, and eventually, the demise of the paddlewheel for boats.

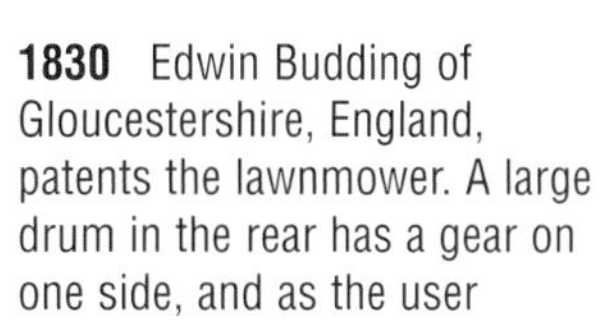

1831 British scientist Michael Faraday confirms induction—metal moving in a magnetic field to generate electricity. Next, he puts wires around a magnet, and forces a suspended disk to spin by continually switching the direction of current in the wires. This is the first electric motor, and the spinning of the disk is really no different than the rotary motion of electric motors used in applications from kitchen blenders to aircraft carriers. (A Faraday electromagnet is mounted under the desk here.)

1830

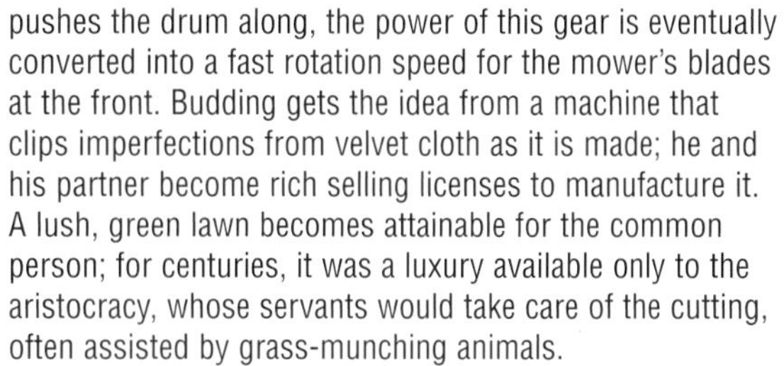

1830 Edwin Budding of Gloucestershire, England, patents the lawnmower. A large drum in the rear has a gear on one side, and as the user pushes the drum along, the power of this gear is eventually converted into a fast rotation speed for the mower's blades at the front. Budding gets the idea from a machine that clips imperfections from velvet cloth as it is made; he and his partner become rich selling licenses to manufacture it. A lush, green lawn becomes attainable for the common person; for centuries, it was a luxury available only to the aristocracy, whose servants would take care of the cutting, often assisted by grass-munching animals.

1830–1870 Americans make significant advances in all facets of rail design, from swiveling wheels to "bells and whistles." The train becomes the vehicle for settling the American West; Golden Spike completes the transcontinental railroad in 1869.

Charging Ahead

It took decades to convert Faraday's wizardry to valuable purposes, but by 1900 huge generators were electrifying the world. Though the process took nearly 75 years, it was a short time compared to the time lag between the invention of previous technologies and their applications, such as gunpowder, which wasn't fully exploited for hundreds of years after its invention. The steamboat and locomotive were successfully deployed 150 years after the power of steam was first harnessed. By the mid-1800s the systematic study and communication of knowledge was increasing the pace of progress, and many more people were now ready to accept advancement.

1831 Cyrus McCormick, son of a farmer, invents a machine that cuts the time needed to harvest a field of grain, from weeks to one day. (Shown, a magazine advertisement from 1875.) The reaper combines a spinning knife with a whirring paddlewheel to scoop up the cut stalks; another paddle knocks the grain off the stalk, and mechanical rakes gather the stalks for baling. America becomes the breadbasket of the world, though the enormous increase in food production comes at the expense of the small-scale farm.

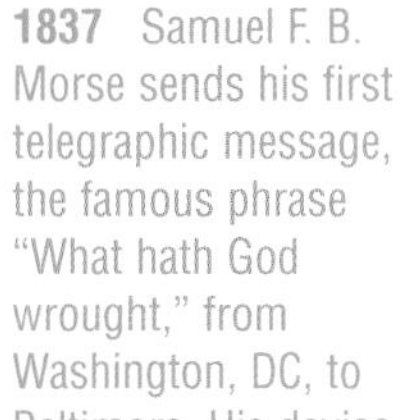

1832 Serving as a sea hand, 16-year-old Samuel Colt studies the locking mechanism for the ship's wheel, and uses his findings to improve the repeating pistol. Onboard, he envisions a pistol that will cock, fire, and feed a new bullet into the chamber in one pull of the trigger. Colt carves a model at sea, and on his return to America, opens his first factory at age 18. That fails, but the boy genius eventually wins world renown, as does the Colt revolver.

1835 The emerging science of the day is demanding ferociously complicated calculations, often carried out by ranks of human assistants called "computers," who introduce a plethora of errors into mathematical tables. So Charles Babbage of England sets out to construct the first mechanical computer, the Difference Engine (shown here), with 25,000 moving parts. His investor, friend, and colleague Ada Byron, Countess of Lovelace, writes the first computer program for one of his proposed machines. Babbage's ideas are sound, but the technology of the day is inadequate to build a functional model.

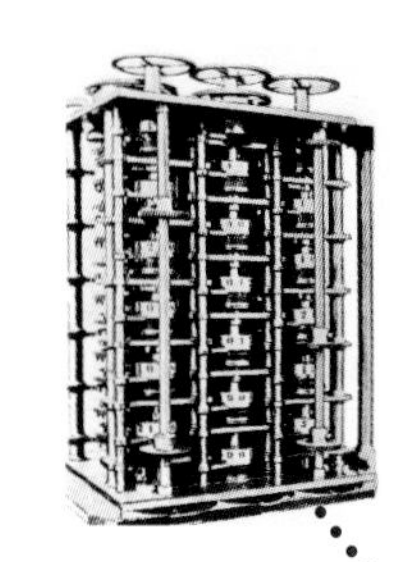

1837 Samuel F. B. Morse sends his first telegraphic message, the famous phrase "What hath God wrought," from Washington, DC, to Baltimore. His device, which owes much to the work of Joseph Henry, is the result of a newfound ability to send pulses of electric current across long distances through switches and relays. For the first time, communication occurs literally at lightning speed. Many lay claim to inventing the telegraph, but few dispute Morse's achievement in devising another great innovation, central to global communications for decades and still useful: Morse code.

Forgotten Greats

The beginning of the Industrial Age produces some of the names we should know, but don't. Some examples:

- **Henry Maudslay** is essential figure in the Machine Age.
- **Joseph Henry** sends signals through wires long before Morse and discovered electrical induction (basis of the generator) earlier than Faraday.
- **William Sturgeon** makes electrical motors workable with his commutator to reverse current flow quickly.

Other geniuses whose inventions are much better known than they are themselves:

- **George Stephenson** (miner's lantern, locomotive)
- **Oliver Evans** (grain elevator, high-pressure steam engine)
- **Jacob Perkins** (nail machine, refrigerator, currency printing plates)

1837 Jacob Perkins patents a workable refrigerator; it uses ether to cool the air, a compressor to recondense the evaporating ether. This system in use today, with Freon or other gas taking the place of ether; no one knows why Perkins didn't pursue manufacturing and marketing more vigorously.

1839 Obsessed with finding a way to make rubber useful in all temperatures, Charles Goodyear, shown here, loses his hardware store and all his money. When he accidentally discovers how to treat rubber with sulfur and baking (*vulcanization* is the word and process he invents), he sends a sample to Englishman Thomas Hancock, hoping for venture capital. Hancock steals the idea and grows rich; Goodyear spends more money promoting his invention than he makes from it, and dies in poverty.

1837 John Deere's "singing plow," named for the hum it makes as it slices the sod, makes the American heartland attractive for settlement. First plow designed from a circular saw blade.

1840 British postal reformer Rowland Hill argues successfully that his nation's system of postal delivery makes no sense: When the letter arrives, the recipient must pay for, and sometimes haggle over, the fee based on weight and distance traveled. Rowland oversees implementation of printed stamps (first one, the "Penny Black" of Great Britain, shown here), to be bought by senders and affixed to the mail as a condition of delivery.

1840 Driven partly by a worldwide soil depletion crisis, German chemist Justus von Liebig discovers the importance of nitrogen as a plant nutrient. After further study, Liebig invents artificial plant fertilizer, founding one of the world's biggest—and most environmentally controversial—industries. Shown here, Liebig in 1845.

1845

Goodyear's Good Spirit

Charles Goodyear's life story is an astonishing cavalcade of bad luck and high honors. One measure of his character: "Life should not be estimated exclusively by the standard of dollars and cents," he writes. "I am not disposed to complain that I have planted and others have gathered the fruits. A man has cause for regret only when he sows and no one reaps." But his descendants—better businesspeople than Charles—prosper from his invention's royalties.

1845 William Morton successfully demonstrates surgery (a tooth extraction) using ether. During the previous year dentist Horace Wells, attempting to show ether's value, accidentally turned the anesthesia off too soon, resulting in an alert and screaming patient. To draw onlookers to the very same amphitheater at Massachusetts General Hospital in Boston, Morton says he'll be using "letheon." He is credited with the invention of modern anesthesiology.

TINGLEY'S PATENT
HORIZONTAL
ICE-CREAM FREEZER
Is recommended for FAMILIES, HOTELS, SALOONS, and WHOLESALE MANUFACTURERS
As the best Ice-Cream Freezer in the market.
It saves ICE,
Saves TIME,
Saves LABOR,
And produces the finest quality of Cream known to the Art.
Send for Descriptive Catalogue.
CHAS. G. BLATCHLEY, Manufacturer,
506 COMMERCE STREET,
Philadelphia, Pa.

1846 Nancy Johnson receives a patent for a hand-cranked ice-cream churn, still used for making ice cream by hand. Johnson sells rights for $200 to William Young, who puts the "Johnson Patent Ice-Cream Freezer" into production. Johnson's churn speeds up the process of stirring cream, flavoring, and other ingredients in a container surrounded by salted ice, the core technique behind making a smooth batch. Shown, 1872 advertisement.

Alfred Nobel

Near the end of a brilliant life of science and invention, Alfred Nobel reads his own obituary in 1888. Published erroneously by a French newspaper (it was Nobel's brother who had died), the headline says: THE MERCHANT OF DEATH IS DEAD. In his will, Nobel starts the foundation that awards annual prizes in his name. The awards reflect his interests (chemistry, physics, literature, physiology) and also his paradoxical hope for peace in a world he unquestionably made more deadly.

1845 William Aspdin in North America makes improvements on a product his father patented in 1824, and modern Portland cement, the world's most common construction material, is born. Consisting of limestone powder, chalk, and clay that is crushed and heated, it's named after stone in Portland, England (shown), famous for its hardness and durability.

1846 Ascanio Sobrero of Italy, experimenting with sulfuric acid and nitric acid, adds glycerin to the mixture, resulting in an explosion. Nitroglycerin is the first significant new explosive in 1,000 years, and when Alfred Nobel (shown here) tries to make it useful, his lab explodes, killing his younger brother. But Nobel persists, finds a stabilizing agent, and calls the resulting product dynamite, which he sells in sticks starting in 1867.

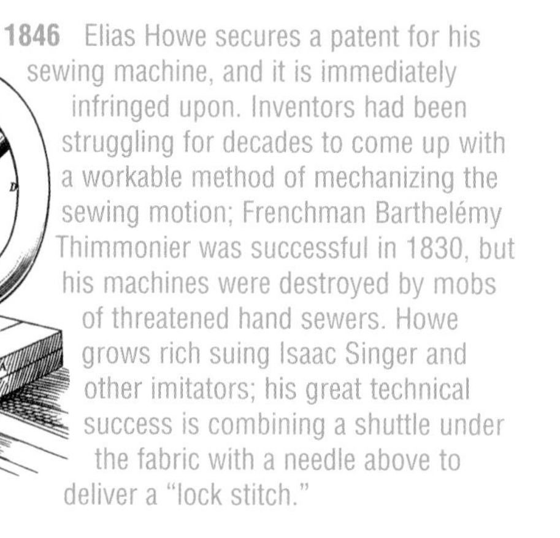

1846 Elias Howe secures a patent for his sewing machine, and it is immediately infringed upon. Inventors had been struggling for decades to come up with a workable method of mechanizing the sewing motion; Frenchman Barthelémy Thimmonier was successful in 1830, but his machines were destroyed by mobs of threatened hand sewers. Howe grows rich suing Isaac Singer and other imitators; his great technical success is combining a shuttle under the fabric with a needle above to deliver a "lock stitch."

1849 Walter Hunt gives the safety pin, an ancient invention, its modern shape. He coils a wire, attaches a fitting to catch and hold its point. The coil, which forces the point much more securely into the clasp than in the traditional design, is the real innovation. Hunt comes up with it after a friend says he'll forgive a $15 debt if Hunt can make something useful with one piece of wire. Seen here, Hunt's patent drawings

1847 Hungarian physician Ignác Semmelweis (shown here on a 1965 stamp) outrages the senior medical staff at the obstetrics ward he heads in Vienna by ordering everyone, including the top doctors, to wash their hands in diluted bleach. Semmelweis doesn't know why, but death rates plummet after the young medical students follow the practice. Senior doctors grudgingly comply but seize on the first chance to have Semmelweis removed. Twenty years later, Joseph Lister popularizes antiseptics, the foundation of modern surgery.

1851 Polite society is shocked and disgusted by Amelia Bloomer's innovation: pants for women. Bloomer, seen here wearing her invention, elicits much ridicule, but "bloomers" prove useful for women—and liberating, later on, with the appearance of the bicycle.

Uniquely Inventive

Abraham Lincoln famously spent time running a flatboat on the Mississippi and Ohio rivers as a young man, and he saw many a boat snagged on sandbars. Lincoln conceived a system of airbags for extra buoyancy, carved a wooden model, and in 1849 received U.S. patent 6469. He is the only president to hold one. Lincoln said the patent system "add[s] the fuel of interest to the fire of genius"; those words are carved above the entrance to the U.S. Department of Commerce building.

1852 Jean Foucault succeeds where others have failed, makes a rotor that stays stable as it whirls around a movable axis. First use of this invention—the gyroscope—proves the rotation of the earth. Eventually fitted into instruments like the artificial horizon in airplanes, the gyro is used to stabilize and navigate vessels and vehicles of all kinds.

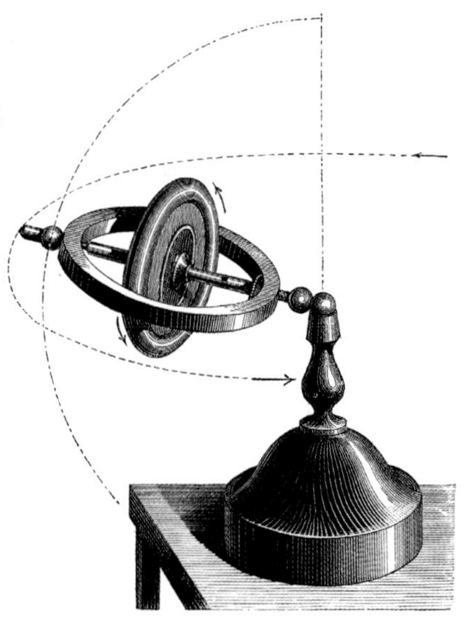

1852 Having seen a worker disabled and an expensive motor destroyed by the shakiness and danger of a lift he supervises, Elisha Otis sets out to make a better version. Adds vertical side bars and wheels for stability and a mesh safety fence, but the real innovation is a spring-released ratchet system on the sides and bottom that brakes the car if the hoist rope snaps. Then Otis invents a name: the elevator.

Rising to the Sky

The Otis elevator improves safety and efficiency, but its real impact is helping to make skyscrapers not only possible but practical. The skyscraper, which changed the design and look of cities worldwide, required cheap but sturdy steel for bearing the enormous weight and stresses of 10 or more stories. First real (all-steel frame) skyscraper is Leiter Building in Chicago in 1889; a 30-year skyscraper boom ensued, transforming cities everywhere.

1856 Proving his critics wrong, Henry Bessemer shows that cool air can be blown directly onto molten iron to efficiently make high-quality steel. Steel had been known for millennia, but Bessemer is the first to develop the hardest grade inexpensively and in large quantities.

Henry Bessemer: Man of Steel

Unlocking the secret of steel was only the highlight of a long career for this engineer's son who, even as a child, spent his days pondering, planning, and producing. Bessemer turned out a cornucopia of inventions, from a ship's berth that swiveled, preventing seasickness; to a new kind of telescope; to a method for solving the bedeviling challenges of getting sugar from sugar cane. And he improves the graphite pencil so that he's sometimes listed as the inventor of the pencil itself. He earned 110 patents.

1853 People have been chewing the bark of willow and silver birch trees for centuries to get pain relief. Chemist Carl Gerhardt, determined to find a way to prevent the searing sores caused by the bark's active agent, salicylic acid, develops a process to buffer the acid, though it's too expensive and laborious for mass production. In 1894, Felix Hoffman comes across Gerhardt's work and brings forth what becomes the world's most commonly used medicine: aspirin. Shown here, a 1950 French advertisement.

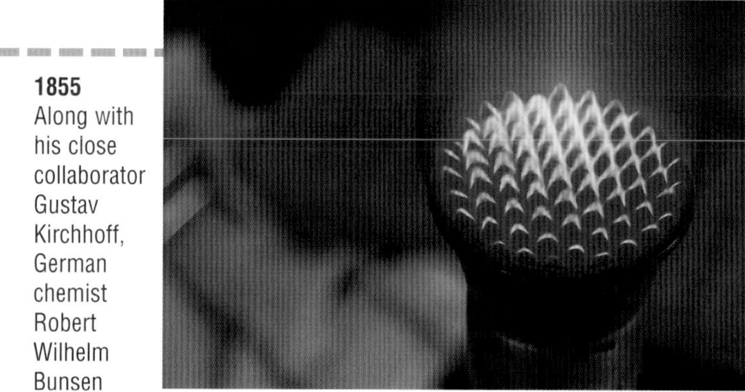

1855 Along with his close collaborator Gustav Kirchhoff, German chemist Robert Wilhelm Bunsen invents the burner that still bears his name and still assists scientists and students every day. A simple metal tube with holes at the bottom delivers a gas flame, controlled by a turnscrew valve. Within a short time, the burner helps Bunsen and Kirchhoff identify two new elements: cesium and rubidium.

1855 Alexander Parkes dissolves pyroxylin, a type of cellulose, in alcohol and camphor. He invents plastic—he calls it Parkesine—but it's so novel that no one knows what to make of it, literally! But soon, a host of inventors are turning out items made of celluloid, the generic name for the product; it doesn't catch on until John Hyatt improves the formula in 1870. It transforms photography when heavy, bulky glass negatives are replaced by celluloid film.

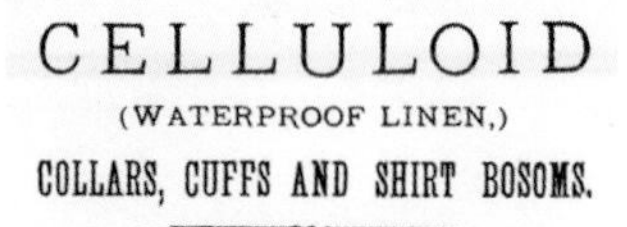

CELLULOID

(WATERPROOF LINEN,)

COLLARS, CUFFS AND SHIRT BOSOMS.

The following will commend the use of these goods to all who study convenience, neatness and economy. The interior is fine linen—The exterior is Celluloid—the union of which combines the strength of Linen with the Water-proof qualities of Celluloid. The trouble and expense of washing is saved.

When soiled simply rub with soap and water (hot or cold) used freely with a stiff brush. They are perspiration proof and are invaluable to travellers, saving all care of laundrying.

ADVICE

In wearing the turn-down Collar, always slip the Neck-tie under the roll. Do not attempt to straighten the fold.

The goods will give better satisfaction if the Separable Sleeve Button and Collar Button is used.

Twist a small rubber elastic or chamois washer around the post of Sleeve Button to prevent possible rattling of Button.

To remove Yellow Stains, which may come from long wearing, use Sapolio, Soap or Saleratus water or Celluline, which latter is a new preparation for cleansing Celluloid.

GOODS FOR SALE BY ALL DEALERS.

DONALDSON BROTHERS, FIVE POINTS, NEW YORK.

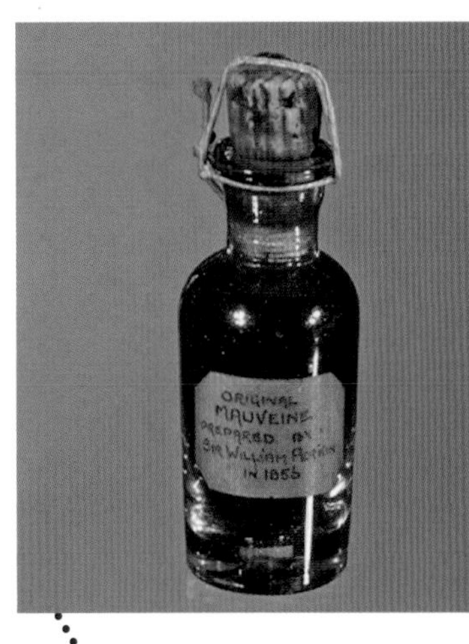

1856 Quinine's usefulness for treating malaria and other problems causes a worldwide shortage. William Henry Perkin, a promising young assistant to renowned chemist A. W. Hoffman, sets out to make synthetic quinine in his home lab. He fails, and then notices the color of the residue from the goo he's accidentally created. It's mauve, the first color from a synthetic dye (seen here), named by a French textile maker. As soon as Perkin sets up a factory to make it, a "mauve craze" sweeps Europe.

1856 A winemaker pleads with the head of sciences at the University of Lille in France: Can't Louis Pasteur, seen here in his lab, figure out the chemical reaction that causes fermentation, and find a way to stop the businessman's sugar-beet wine from souring? Pasteur does more than that: He determines that yeast (which causes fermentation) is a living organism, not a chemical, and it—along with a host of disease agents—can be killed by heat. Food processing and public health are transformed, saving untold numbers of lives.

1858 New York City inventor John L. Mason devises a shaping process that puts screw threads on the inside of metal jar tops. Matched with the threads on a glass jar, and fitted with a rubber seal and zinc lid, his invention makes safe canning practical in the home; Mason jars become ubiquitous.

1859 Steam engines use external combustion; that is, the gas to move the pistons is heated from the outside. But engines could be smaller and more useful if the gas is combusted inside the engine. Belgian Étienne Lenoir's engine, which ignites coal gas with a spark, really isn't any smaller or more efficient than the existing steam engine, but it's the first engine using internal combustion.

1859 Edwin Drake invents the oil well. He wonders if oil could be drilled for, like water, and is the first to try it out successfully, in Titusville, Pennsylvania. The Drake Well was immediately and widely copied.

1850s Cigarettes become the rage across the Continent after an unfashionable start: They are rolled by street people in Seville, Spain, from chopped-up cigar butts. *Cigarette* is French for "little cigar." Shops begin carrying hand-rolled cigarettes in 1860s; James Bonsack invents cigarette-making machine in 1880. Vincent Van Gogh unintentionally captures a consequence of smoking in 1885, with *Skull with Cigarette*, seen here.

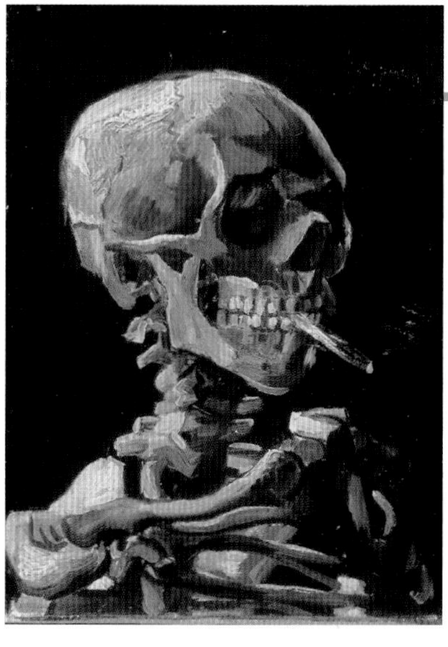

1860 Frederick Walton of Britain is the first to develop a smooth, tough, cheap floor covering that's easily installed. He devises a mixture of resin, gums, and linseed oil that, when coated on burlap or felt, makes linoleum. It is the floor covering of choice for many decades, till replaced by vinyl.

1861 Linus Yale Jr. and Sr. become synonymous with their invention, the "pin tumbler" lock. Yale Sr. had arranged spring-loaded pins that allowed lock to turn only if matching key is inserted into it. Yale Jr. improves device with cylinder mechanism and flat key still seen today, often stamped with the name Yale, making the whole thing more compact and secure.

Invention's Toll

The American Civil War makes it possible to measure "progress" in the worst way imaginable: by counting the numbers of dead and wounded. The Industrial Revolution's technology has far outpaced the values and tactics of war strategists. It's still considered cowardly to break ranks and seek cover, and "generals throughout the Civil War sent their men marching in great masses toward the enemy over open fields, as had been done in the American Revolution and virtually every war in Europe," to quote one war historian. The result: Modern, repeating rifles, machine-tooled revolvers, and giant new cannon mow down ranks of men in huge waves. In the war's aftermath, a whole new approach to fighting emerges as tacticians realize that what they've inflicted on the troops is "simply murder," as one Confederate soldier puts it.

1862 Richard Jordan Gatling unveils the first functional machine gun; the idea had been around since Leonardo conceived a rapid-firing crossbow but defied efforts to make it functional. The Gatling gun uses a hand crank and fires 250 shots per minute, and Union Gen. Benjamin Butler orders several. The guns perform well, but the army's first large order is placed in 1866, after the Civil War is over.

1866 Shirt maker Oliver Winchester, having invented the striped shirt with white collar and cuffs named after him, puts his earnings into the Volcanic Repeating Rifle Company, run by the makers of the famous Smith & Wesson. He hires Benjamin Tyler Henry to improve the product, and the "Henry rifle" is the first widely praised and popular rifle that can fire without reloading between shots. The first Winchester rifle is the Model 1866, and is known as "the gun that won the West."

1868 Reporter Christopher Sholes of the *Milwaukee News* has an unoriginal idea: the typewriter. But he has a few novel ways to solve problems that caused earlier renditions of the machine to fail. He gets the ideas for the keys from watching a friend play piano, and invents the "QWERTY" keyboard to keep commonly used letters spaced apart so they won't jam. Sholes's first models flop, but he perseveres, creating a new industry and bringing women into the business workplace in unprecedented numbers.

1869 France's Napoleon III has put out a call for an alternative to butter, and H. Mège-Mouriès responds with a mixture of skim milk, suet, and the innards of pigs and sheep, which he calls *butterine*. It's a much bigger hit than one might expect, given the ingredients list. It's also the source of outrage from the dairy industry worldwide, whose lobbyists in Britain impel the government to require a name change. Mouriès turns to the French word for animal or vegetable oil, and coins *margarine*.

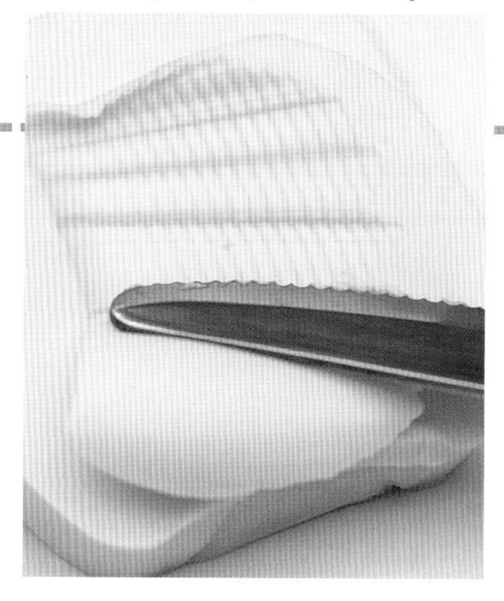

1868 In 1866, George Westinghouse (shown) was delayed on a train trip because of a collision elsewhere on the line. He recognized a problem and opportunity: Many train accidents might be avoided if a braking system could be devised to stop all the wheels at once. Inspired by an air-compression system used to tunnel through the Swiss Alps, he invents a revolutionary braking system that speeds up train travel by giving operators more confidence in their ability to slow down.

1869 Thomas Adams, an American photographer working in Mexico, buys some chicle, a gum Mexicans love to chew, derived from a species of evergreen. Adams suspects it would be popular in America, and he's right. After he adds flavoring and convinces a local druggist to sell it, his gum becomes one of America's favorite guilty pleasures.

1870

1870 Thomas Alva Edison sells his first successful invention, the Universal Stock Ticker (shown). It is actually not the first stock ticker—that machine, whose ticking telegraph keys inspires its name, was invented in 1867 by Edward Calahan of American Telegraph Co. But Edison's is a big improvement and a bigger success than his first invention, an electric vote counter for legislative bodies (1868).

1873 First effective barbed wire invented by Lucien Smith, 1867; after seeing it demonstrated at the DeKalb, Illinois County Fair, Joseph Glidden and his friend Jacob Haish both quietly try to become the first to patent a method to make the stuff in quantity. Glidden wins. First twists barbs successfully with his wife's meat grinder and soon opens a factory in DeKalb. Stringing of barbed wire opens the "range wars" between Western ranchers and farmers; their opposing needs collide violently for decades.

Edison

Thomas Alva Edison was born February 11, 1847, in Milan, Ohio; he had only a few months' formal education but becomes a lifelong learner and believer in continuous self-improvement. He grows up in Port Huron, Michigan, and is homeschooled by his mother, whom he credits for his success. He receives 1,093 patents solely or jointly, by far the all-time record holder. News of his death October 18, 1931, prompted President Herbert Hoover to ask Americans to dim their lights in Edison's honor at 10 PM the next evening. It was even suggested that the electrical grid itself be turned off briefly, but doing so would have dangerously disrupted facilities such as pumping stations and hospitals. "This demonstration of the dependence of the country on electrical current for its life and health is in itself a monument to Mr. Edison's genius," Hoover said.

1876 Anna Bissell is at wit's end trying to sweep the sawdust from the carpet of the crockery shop she owns with her husband, Melville. She implores him to devise a better sweeper. He attaches two wheeled brushes that whirl dirt into hollow base. On Melville's death in 1889, Anna becomes first American female CEO of major corporation. The Bissell carpet sweeper still sells well today.

1875 With completion of the London sewer system, Joseph Bazalgette saves countless lives, perhaps Western civilization. Plagues of fever-producing killer diseases like cholera routinely ravage European cities as the sewage of crowded urban populations sits fetid in the streets, lowlands, and lifeless rivers. Driven from London by the "Great Stink" of 1858, Parliament appoints civil engineer Bazalgette to find a solution. The result is the Thames Embankment (seen here). When other cities are blitzed by cholera in the 1870s, London is spared—and Bazalgette goes from being ridiculed to being knighted.

1876 Alexander Graham Bell and Elisha Gray file independent applications to patent the telephone. Bell files "Improvements in Telegraphy" even though he hasn't yet been able to complete a communication with his telephone. But on March 10, 1876, having spilled some acid in his Boston lab, Bell yells, "Mr. Watson, come here, I want you!" and a new age in human interaction is underway. Because Bell's original design would not have worked and Gray's did, credit for the invention of the telephone is still disputed.

Edison's Best Idea?

The phonograph. The motion picture. The electric light. None of these was Edison's greatest invention, in the minds of many technology historians. In 1876, Edison consolidated his earnings and invested them in a shop equipped with sufficient tools and materials to investigate or invent almost anything. It is a brand-new idea: the research and development laboratory. Edison chooses Menlo Park, New Jersey, about 25 miles southwest of New York City, as its site, and the concept's value is soon obvious: Edison becomes known as "the wizard of Menlo Park."

1877 As with the telephone, the phonograph is an accidental offshoot of telegraphy. Thomas Edison is looking for a way to record Morse-code pulses; he and his team mount a wired stylus on a cylinder, and pulses are impressed in tin foil. After some refinement, a mouthpiece is attached to record vibrations of sound. A speaker can be substituted for the mouthpiece. "Mary Had a Little Lamb," recited by Edison, is the first sound recording. Edison, the most prolific patent holder in history, calls the phonograph, pictured here, his favorite invention.

1879 Electricity heats up metal when it flows through it, causing a bright glow. That's well established by 1815, but it takes 50 more years to create a useful electric light—the primary problem: finding a tough-enough filament and a bulb that keeps oxygen out and the filament from burning up. Edison tries an estimated 10,000 times, and Englishman Benjamin Swan, the "other" inventor of lightbulb, even longer. They both arrive at success at the same time. Swan's solution is on the left; Edison's, on the right.

1879 Fannie Farmer, a "mother's helper" in Boston, attends a lecture by Catherine Beecher expounding on the need for household work to become as scientific as the rest of the world. Farmer researches cooking and invents new, precise, and practical measures like the cup, tablespoon, and teaspoon and puts a thermometer in a cookstove for the first time. Farmer's recipes end 40 millennia of hit-or-miss cookery; her units have undergone no important modification.

1883 Serbian genius Nikola Tesla, photographed here with an oscillator, invents the electrical motor as we know it by figuring out a way to make a stationary wire coil deliver a rotating magnetic field. Motor relies on another Tesla miracle creation: the "three-phase" electrical supply, which delivers more than one voltage through the same set of wires. Huge increase in efficiency. He promotes his ideas in America.

1884 James Ritty, a saloon owner in Dayton, Ohio, is losing too much money to dishonest bartenders. He rigs a cashbox to numbered keys connected to a big dial that indicates amount tendered and keeps a total of what should be in the till. Gets the idea from ship's engine room, where he sees a dial that counts the turns of the propeller. Ritty sells patent rights to James Patterson, who founds National Cash Register.

AC/DC

Edison and Tesla waged one of the great feuds in the history of science—especially noteworthy because Edison took the losing point of view. Edison instituted the first large public electric supply in history, in New York in 1882. But it relied on direct current, which could not travel long distances and caused sparks. Tesla had come to the U.S. having developed a better, safer alternating current (back-and-forth) system of power generation, Tesla's former boss wrote Edison in a letter of reference, "I know of only two great men. You are one of them, and the other is this young man." But this dynamic duo was not destined to fulfill its potential. Tesla worked for Edison, but could not convince him of the wisdom of switching to AC. They quarreled over royalties for Tesla's brilliant work and parted as enemies. Tesla formed his own company and lost it, and became a ditch digger. Eventually, though, he teamed up with George Westinghouse, of air-brake fame, and the pair wire the world for AC. Tesla grew more eccentric as he aged, went bankrupt funding his overly ambitious projects, and eventually lost the support of Westinghouse and his other backers. He died poor and surprisingly unknown in 1943.

I Want to Ride My Bicycle

It was the bicycle that really taught the world about individual power and freedom that is commonly associated with the automobile. Bikes were relatively cheap and made fast transportation a matter of personal choice in a way nothing else had. Through the 1890s, people everywhere discovered true personal mobility; couples from adjoining rural villages could meet up on their own schedules for the first time, to name but one social impact. Another was a huge shift in women's attitudes: the bicycle made "bloomers" (pants) fashionable, not shameful, and suggest to a generation of women they have a right to exercise both their bodies and wills. As technologies of all sorts raised the standard of living and opened up more leisure time, people of all classes learned to spend time playing as a daily part of life, and the bicycle became a significant element in this evolution.

1884 William Herschel and Henry Faulds determined in 1880 that fingerprints never change. Now, British anthropologist Francis Galton proposes collecting and keeping databases of fingerprint records for identification. In 1893, British committee declares conclusively that no 2 fingerprints are alike, establishing their admissibility in court.

1884 Fully automatic machine gun is invented by Hiram Maxim. Uses recoil action to feed cartridges from a long belt. Use in WWI inflicts unprecedented casualties and ends the tactic of the mass charge by infantrymen. Maxim (pictured here with his grandson) is knighted.

1885 William Burroughs unveils the "adder-lister," the first adding machine that prints out its calculations. It takes 7 years of improvements before it makes a major impact, but after 1892, it becomes mandatory for well-run financial establishments.

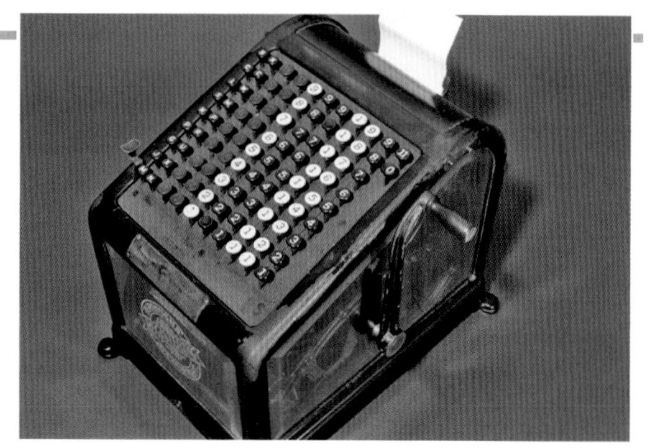

1885 British engineer John Starley consolidates a number of innovations to create the first "safety," i.e., modern-looking, bicycle. Both wheels are about the same size; it's efficiently chain-driven; and, in editions shortly after the first model, has the distinctive diamond-shaped frame and curved front forks. In 1888 John Dunlop comes out with the smooth, fast-running pneumatic tire, completing the package.

1885 Two vital and fast-developing technologies merge to create the first motorcycle. Gottlieb Daimler wants to test his gas engine, and bicycles are maturing into tough, well-balanced vehicles, so he does the next logical thing: rigs the cylinders of his engine to the rear wheels of a modified bike. The intention is solely to test the engine, but he actually creates an entire class of vehicles, an industry, and eventually a lifestyle.

1885 By now, the slow and inefficient Lenoir engine of 1859 is greatly improved. New, more explosive fuels are being contemplated to turn the "four stroke" engine invented by Nikolas Otto into a power plant that will make fast cars and even airplanes possible. Karl Benz and Gottlieb Daimler come out with a gasoline engine that relies on a new element, the carburetor, which turns liquid fuel (anything volatile will work) and air into a mist for combustion.

1885–1891 In a blazing demonstration of how quickly technical evolution is now taking place, the automobile goes from a 3-wheel, single-cylinder prototype to clutch-shifted, rear-wheel-driven city vehicle in 5 years. German engineer Karl Benz builds the first one; Gottlieb Daimler and others add gears and the 4th wheel; and Frenchman Émile Levassor puts the engine in the front.

1888 George Eastman changes photography into a medium for the common person by inventing what he calls the Kodak camera. "You press the button, we do the rest" is the Eastman Kodak slogan; customers buy a preloaded camera, shoot the film, and ship it to the Kodak plant. It comes back with prints and a fresh roll of film. Among the many impacts: Americans are photographed smiling, which seldom happened with daguerreotypes.

The Kodak Camera.

"You press the button, we do the rest"

(OR YOU CAN DO IT YOURSELF).

The only camera that anybody can use without instructions. Send for the Primer, free.

The Kodak is for sale by all Photo stock dealers.

The Eastman Dry Plate and Film Co.

Price, $25.00—Loaded for 100 Pictures. Reloading, $2.00. ROCHESTER, N. Y.

379.18 THE KODAK #1, 1889.
Credit: The Granger Collection, New York

1891 Having perfected the safety elevator, the Otis corporation takes the next step: a chain of moving steps; actually, Otis acquires rights to the inventions of Jesse Reno, who created a novelty escalator at Coney Island, and Charles Seeberger, who patented and copyrighted the escalator. Otis installs its first escalator in its own factory; public models are in operation in the New York City subway and the Paris Exhibition by 1900.

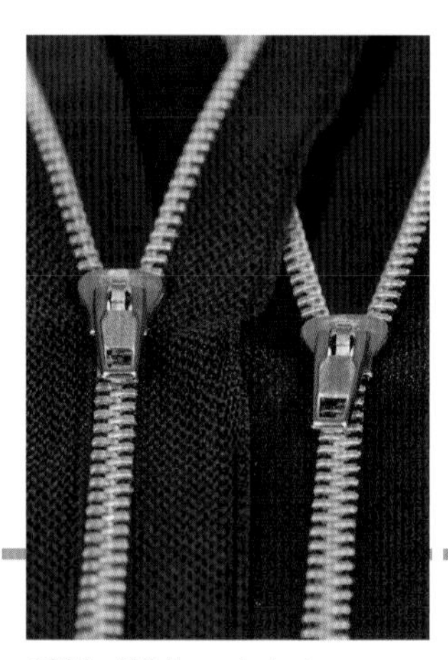

1891 Whitcomb Judson, a mechanical engineer in Chicago, is fed up stooping to button his shoes and plots a better fastening system. Designs a system of hooks that mesh with guidance of a sliding tab. Gideon Sundback of Sweden makes the new idea much more functional; and in 1923, BFGoodrich incorporates the fastener in a new line of chic boots with a catchy name: the zipper.

Patent Medicines

Before the colonies were even settled, so-called patent medicines had appeared in England. Actually, true patent medicines were probably the most respectable kind —which isn't saying much. The name came from the British patent requirement: The medicine's inventor had to prove an original formula, not the medicine's effectiveness or safety. Neither adjective applied to 99 percent of the quack medicines that abounded in the 19th century. There was a patent medicine for every malady from cancer to baldness, and some claimed to cure everything. The ferocious promotion of the endless varieties of patent medicines spurred the development of many modern techniques of advertising and sales. But even here the industry's legacy is shady: "snake oil salesman" is still used to mean con artist. It stems from just one of the endless number of miracle ingredients the concoctors claimed were cure-alls. The business is driven by justified terror of disease; most of it withered with the advent of modern medical practices, though a look through today's supermarket tabloids demonstrates it hasn't entirely disappeared.

1892 The age of agricultural mechanization begins in earnest when Iowa blacksmith John Froehlich invents the gas-powered tractor. Soon results in a major increase in productivity, standard of living, and population increases so pronounced that overpopulation becomes a major concern.

1895 Having spent the past year underwhelming people with his newly developed ability to send electromagnetic signals a tiny distance through the air, Guglielmo Marconi finally gets everyone's attention by finding an antenna design that can pick up radio signals from more than a mile away. Birth of radio and wireless communication, and within a decade, ships have radio rooms and the first voice broadcast is made.

1893 Picking up on a famous 1877 experiment that proved all four hooves leave the ground during part of a horse's stride, Thomas Edison and assistant William Dickson invent a camera that captures images in quick sequence, and a projector to show the film. His "kinetograph" is a sensation, but is viewed by just one person at a time; in 1895, Frenchman Antoine Lumière and his sons turn the kinetoscope into a moviemaking camera by inventing a truly portable unit and a projector for large-scale viewing.

1898 With the telephone fast becoming a household item, Valdemar Poulsen sets out to create an answering machine—and invents something far more significant. Poulsen's solution is to coat wire with iron oxide (rust), which captures a magnetic imprint of sound created by the electric pulses of a microphone. This concept—of storing pulses on magnetic media—proves to be one of the most useful and important technologies, in video and audio tape, computer disks, stripe readers, and more.

Also-rans

If only every invention were as useful and elegant as the paper clip. Wishful thinking: While historical timelines tend to feature the breakthroughs and triumphs, the truth is that most patented inventions fail. And quite a few raise the fundamental question, "What were they thinking?" Take the automatic egg counter, to be worn on a chicken's rump. Or the bunny-shaped hypodermic needle, for pediatric injections. Or the auto powered by a dog on a treadmill. These and many, many more like them received patents. A small but fascinating genre of technology scholarship is devoted to the long history of goofy, dangerous, or lamentable flashes of inspiration. In 1973, researcher Stacy V. Jones penned a memorable title for his compendia of forgettable ideas: *Inventions Necessity Is Not the Mother Of.*

1899 Johan Vaaler of Norway invents a superior way to keep stacks of paper together. But because Norway has no patent system, he has to go to Germany to secure rights to the paper clip. In his 1901 patent application Vaaler proposes various different ways of bending the wire (in triangular or square shapes, etc.). In the meantime, the Gem company of Britain invents the classic paper clip shape (the "Roman"); no one has ever patented it.

1900

1900 German Ferdinand Graf von Zeppelin unveils the aluminum-framed hydrogen airship that bears his name. It actually represents the pinnacle of a series of such rigid lighter-than-air devices, many successful, developed in increments during the last part of the 19th century. The zeppelin is copied by the British, and industry is thriving when the *Hindenburg* horrifies the world by bursting into flames, killing 36, as it docks in Lakehurst, New Jersey, in 1937. The *Hindenburg* ends the hydrogen era; some think the safer ships of today represent an evolving future technology.

1901 Around the turn of the 20th century, so it's said, a friend advises ambitious businessman and tinkerer King Gillette to invent something people throw away and must keep replacing. He's thinking this idea over while shaving, and the answer is right in front of him. A century later, Procter & Gamble pays $57 billion to acquire the Gillette Company.

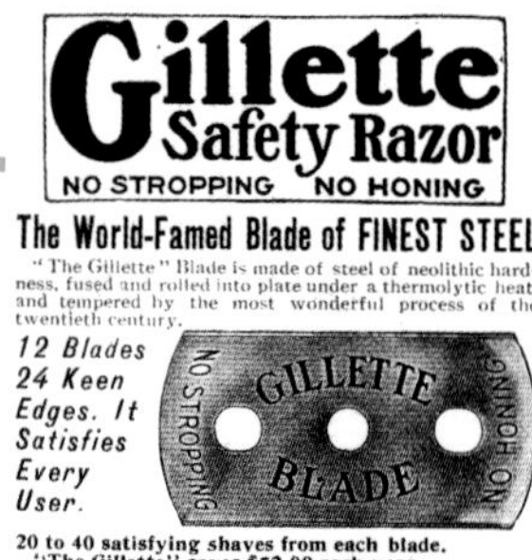

566.25 GILLETTE RAZOR AD, 1906.

1901 Hubert Booth sees a train cleaning crew using a device that blasts dust out of upholstery with compressed air—and then lets it settle right back again, without collecting it. He has the defining reaction of the true inventor: "There's got to be a better way." The first vacuum cleaner is a 2-person contraption: a 5-horsepower engine mounted on a wagon pulls up in front of a house; one man operates the engine, and the other runs the suction hose in through a window and cleans the carpets.

Liftoff

The Wrights overcame three problems that defeated others. Their mechanic devised perhaps the best gasoline engine yet. Their "wing warping" system successfully addressed the issue of control in the air. And they realized the propellers should simply be wings turned vertically. The universal key to finding these solutions was a methodical reexamination of the "facts," such as mathematical tables, underlying the conventional wisdom about flight. Wilbur later recalled hte low point of their quest, in 1901, "We considered our experiments a failure. At that time I made the prediction that men would sometime fly, but that it would not be within our lifetime."

1902 Willis Carrier's friend asks him to take on a vexing problem, and Carrier's solution leads to the air conditioner. The friend's color printing business is going badly because paper shrinks slightly at higher temperatures. Carrier uses expanding ammonia forced through tubes to create super-cold air; water from the cooling room air condenses onto thin metal baffles joined to the tubes, dehumidifying the air.

1903 Succeeding where famous and well-funded scientists had long failed, 2 bicycle makers from Dayton, Ohio, redefine the meaning of "possible" at 10:35 AM on December 17, when their heavier-than-air airship lifts off from the sand at Kill Devil Hills, North Carolina. Orville Wright is the pilot and brother Wilbur runs alongside in what might be the most widely reproduced photo ever. Total investment to realize this millennia-long desire: about $1,000.

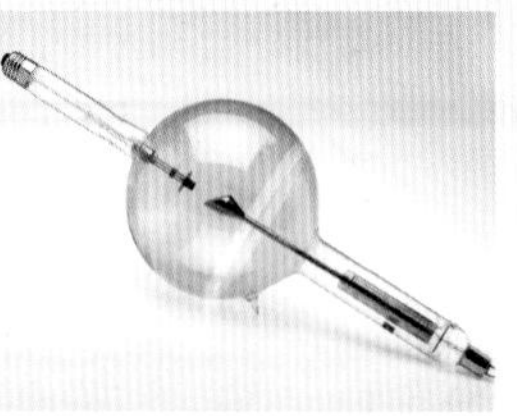

From the Airplane to the Internet (and the Future)

Once the Wright brothers had destroyed the old definition of *impossible*, the impossible was accomplished at an increasingly rapid pace. That's really what the last century has been about: inventors achieving the impossible, the unimaginable, the unthinkable—and sadly, the unspeakable.

While people had been debating the viability of powered flight long before the Wright brothers left the ground, many achievements of the twentieth century—like cloning, handheld computers, and nuclear bombs—didn't spark the same kind of arguments beforehand, because they were too outlandish to make for good debate. Certainly it never occurred to anyone that artificial satellites would be circling the earth barely fifty years after Orville Wright astounded humanity by simply leaving the ground for twelve seconds.

But the burgeoning of an elite scientific corps, aided by better and better mathematical tools and physical instruments, meant mysteries were solved at an ever-increasing pace—faster than anyone expected. And each new wonder laid the foundation for something even more incredible, as in the evolution from radio to television, or walkie-talkies to cell phones. Science led the way, and engineering followed up, with tinkering and discovery occurring at the molecular, then the atomic, then the subatomic level. The state of technology moved into realms beyond what people had been dreaming about for millennia, to levels more amazing than dreams—beyond the solar system, into the DNA of dinosaurs.

Math and physics were the foundations of the century's astounding work, beginning with the Wright brothers. Their fundamental key to unlocking the secret of flight was revising the mathematical tables describing wing area and lift, and putting them on the path to a workable design. Similarly, rocket pioneer Robert Goddard is remembered in the public mind standing next to one of his early liquid rockets, but his most important achievements weren't physical. They were scribbled on the blackboards of Worcester Polytechnic Institute, or posited, corrected, and refined in his hundreds of notebooks.

Communications was the same way: What made radio, television, cable and satellite entertainment, computer networks, wireless computing, and digital telephones possible, fundamentally, was a mathematics sophisticated enough to describe interplanetary signal transmission or the flows of current in a transistor one molecule wide. Getting the numbers right made the digital world possible—*digital*, after all, means numeric.

This was true of all the myriad tools, toys, devices, and weapons rendered by modern physics in the century. And the mathematics did not have to be complicated to be profound—just elegant, as exemplified by the most famous algebraic expression since Pythagoras: $E=mc^2$. Einstein's beautiful and deadly evocation of the relationship between matter and energy is the ultimate example of the potential for math and physics to affect the world.

But, actually, it is one example of the thousands of ways that experimentation with the physics of the atom created the age of nanoelectronics, space travel, and gene therapy. Through a combination of math and experiment, physicists came to understand how electrons, protons, and photons behave; what magnetism can do to them; and what they can be made to do to one another. This was the most important science behind the thrilling and frightening century that launched the postmodern era.

The collaboration between mathematicians and physicists was most evident in the products of nuclear physics—bombs, power plants, ship engines—but just as applicable and important in the creation of the microwave or the computer chip or the magnetic resonance imager (MRI). From greeting cards that play electronic "Happy Birthday" tunes to cameras that transmit images from deep inside the body to telescopes that detect the echo from the creation of the universe, much of the advancement of the last century derives from the convergence of math, physics, and chemistry at an infinitesimal scale.

The advance of technology affected everything but may have been felt most profoundly in the life sciences. When the century started, people were still helpless against most diseases, though vaccination had been invented and sanitation and hygiene had come a very long way in the previous fifty years. Women lived shorter lives than men on average because of the dangers of childbirth. Yet within a century, the average Western lifespan had gone from forty-five to seventy-five years, and the quality of those older years had increased apace. This amazing improvement happened one practical invention or innovation at time—a drug, a device, or a technique.

While invention changed the world, the craft itself changed greatly too. It was possible to build the Wright Flyer in 1903 without a college education, but Goddard could not have built the first liquid-fueled rocket in 1926 without going to the polytechnic institute and university. The math and theory behind noteworthy accomplishments in engineering palpably moved out of the realm of the self-taught and the trial-and-error schools, especially in the latter half of the century.

Innovation also moved past the province of the lone genius. Many achievements, even today in the age of the lone computer "geek," are personified by the celebrity entrepreneurs who promote them, but the reality is that life-changing breakthroughs like the Internet or Prozac are far more often the result of teamwork than earlier milestones like the microscope or telegraph. And these teams are far more formally organized, in the research-and-development departments of huge capitalized corporations like Sony and Raytheon and Boeing. It was teams of top scientists who invented the atomic bomb and the transistor; Edisons and Franklins were always rare and are becoming even rarer as the pace of invention accelerates.

But the quest to improve, solve problems, and startle the world is exactly the same now as in 1900, 0 AD, or 3400 BC. At the end of a million years of ingenuity, humankind finds itself where it started: the beginning of the future.

1904 Thomas Sullivan, a tea seller in New York, invents the tea bag—or rather, his customers do. He distributes hundreds of samples in innovative fashion, mailing or handing out cloth pouches of tea. Rather than opening them, recipients just dunk them in hot water. Development of a bag that doesn't flavor the tea takes decades. First mass-produced teabag is introduced in Britain in 1953 by Joseph Tetley, and bag sales really take off in America during the 1960s with Lipton's flow-through bag.

1900

1905 On a trip to New York, Alabaman Mary Anderson watches her driver fight to keep the windshield clear of snow and ice. Anderson devises a hand-operated wiper arm for the window, connected to a lever on the dashboard. The windshield wiper is born.

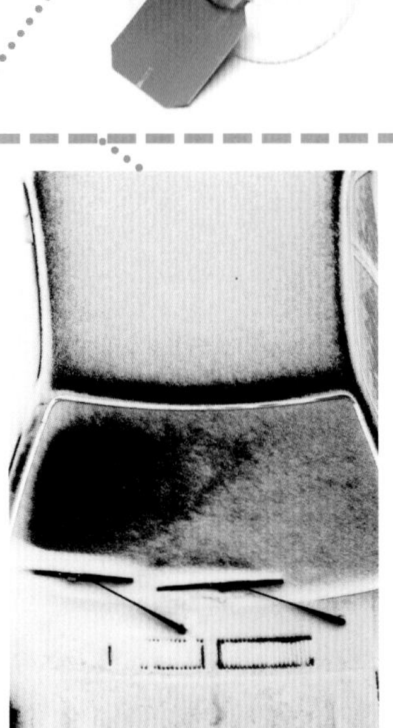

1906 Out of high school and bored working at his father's hardware store, Alva Fisher decides he'd like to become an inventor, but can't decide what needs inventing. His mother says it should be something to help her around the house. Fisher sets about rigging her hand washing machine to a motor, mechanizing its functions one by one. Patents the automatic washer in 1910, ending eons of backbreaking drudgery (as seen in this 1929 ad) for women and providing them a huge amount of new free time.

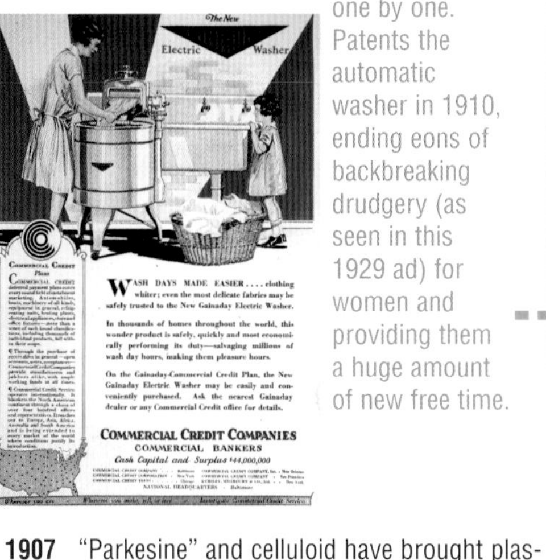

1907 Brothers Auguste and Louis Lumière develop workable method of color photography. They mix potato starch in red, blue, and green ink; the starch granules act as color filters to represent true colors on the final film, which is projected like slides. Eventually film is treated with chemicals instead of starch.

1907 "Parkesine" and celluloid have brought plastic to the world, but now Leo Baekeland of Belgium, a chemist, takes the step that will bring the world to plastic via a strong, light, heat resistant, and easy to cut and shape material. Much of the manufactured world is now re-created using Baekeland's invention (Bakelite, a lidded bowl shown here) and its successors (PVC, polyethylene, Teflon, etc.). Within a few years, thousands of Bakelite-made products appear along with a massive new environmental problem.

1908 The gyroscope and its possibilities have fascinated scientists ever since its invention by Leon Foucault. Elmer Sperry now makes a compass that exploits the gyroscope's tendency to return to true north if placed in a housing that shifts along with the earth's rotation. Era of magnetic compass ends. Besides use on ships, gyroscopes are soon placed in airplanes as autopilots. Today's finest gyroscopes use light beams, not wheels.

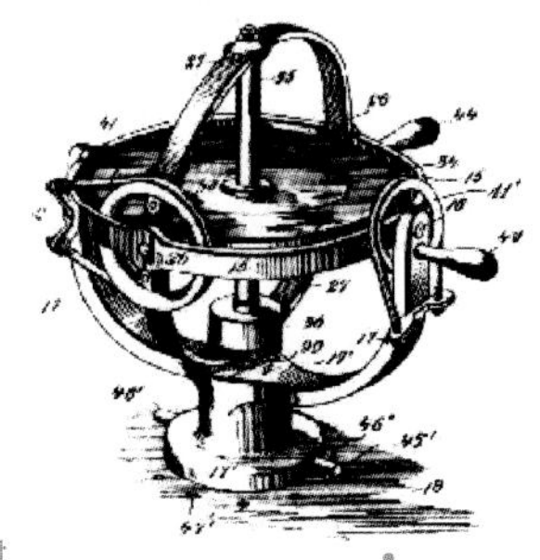

Better Living

After 1870, there was an obvious shift in the focus of invention. The process of converting discovery into practical application continued as it had for millennia; As new inventions became old hat, the desire to improve on them drove the next round of innovation: questioning, research, application. But in the late 19th century and later, the fruits of science are increasingly applied to products for the consumer market; the laboratory becomes the source of a constant stream of creature comforts. And earlier bursts of automation, industrialization and electrification, meant efficiency had risen so much that ordinary people had more leisure time to enjoy their new consumer goods, and more disposable income to buy them. The middle class was growing, and the consumer society was being invented.

1912 English metallurgist Harry Brearley searches for a way to keep steel cannon barrels from corroding. One day he's hunting through a pile of scrap metal and notices a piece that's still shiny. Investigating its composition, he determines that nickel and chromium steel, combined properly, won't rust or stain. Though stainless steel never does make a corrosion-free gun barrel, it does become one of the world's most used and useful manufacturing materials in everything from battleships to artificial hips. First use is in cutlery, where it completely changes the laborious process of cleaning after each meal.

1913 On December 21, the New York *World* debuts a new feature: the crossword puzzle, invented by British staffer Arthur Wynne at his editor's request. First-ever published crossword clue: "What bargain hunters enjoy."

1913 William Coolidge of MIT invents a cathode-ray tube specifically for the creation of X-ray images for medicine. Discovered in 1895, X-rays pass through flesh to provide images of what lies beneath, with different densities revealed as shades of gray. The new tube is so useful, it becomes known as the "Coolidge tube" in honor of its inventor, and has not required profound modification in the century since.

Amazing Rays

X-ray machines for medicine were just one example of the fantastic and incredibly useful 20th-century by-products of the development of vacuum tubes, which began in the 1850s. Greats such as Michael Faraday investigated the flow of current through these tubes—airless glass chambers with a positive contact (anode) and negative contact (cathode) placed at either end. As everything about them improved—glass, wiring, completeness of vacuum—the tubes grew more powerful and reliable, both in generating the current between the two electrodes (expressed as a bright glow) and causing objects placed in the path of the beam to heat up (such as the platinum strip labeled *b* in this 1879 rendering). As scientists explored the effect of magnets on the behavior of the current in the tube and new experiments were conducted, it became clear that current is a stream of particles, eventually called *cathode rays*. J. J. Thomason of Britain won the first Nobel Prize in Physics (1901) for identifying the particles as electrons, the negatively charged component of the atom.

Scientists and inventors set out to find ways to create and manipulate streams of electrons and associated radiation, using variations on the basic cathode ray tube (CRT). In 1895 Wilhelm Röntgen accidentally discovered X-rays by observing a container of barium platinocyanide on a shelf in his lab. It glowed when he turned on a CRT in a box. He soon astounded himself and the world by determining these new rays (they were energy packets, not particles) could go though wood, paper, clothing, and flesh. X-ray machines now abound in settings from the factory to the airport security line to the dentist's office. The neon light and the fluorescent lamp are both consequences of the CRT. In 1907 Boris Rosing, with ideas that outpaced the technology of his time, described how a CRT could produce an image if it has an optically sensitive surface at one end. His proposition: television.

In 1927 Philo T. Farnsworth made Rosing's idea a reality. And the miracle of the CRT didn't end with TV; from air-traffic control screens to cash registers to computer monitors, the CRT had changed daily life, mostly for the better, for another 100 years with no end in sight. And it affected not just the family room or the mall. CRT imaging detects objects so small, they escape the wavelengths of the visible light spectrum and the naked eye. In the minute wavelengths produced by CRTs, the same objects readily appear, making the electron microscope one of the most important developments of the 20th century. And in the 21st century nanotechnology, the practical application of the science of the incredibly minute becomes one of the most promising cutting-edge scientific disciplines. Founded on the electron microscope and its successors, nanotechnology seems certain to yield outcomes as astounding as the first X-ray images were in 1895.

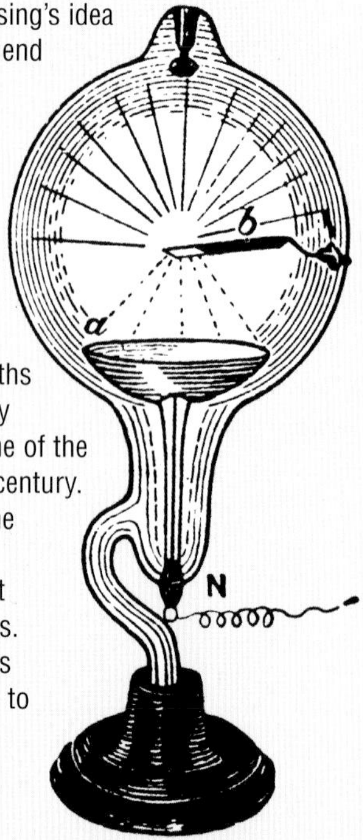

1913 Henry Ford is unhappy even though he's selling more than 80,000 Model T's a year. He believes he could take the automobile from handmade luxury item to mass-produced vehicle for average Americans if he could find a way to get production costs down. After a 1912 visit to a slaughterhouse, where meat is processed by a string of stationary workers in front of a conveyor belt, Ford implements the same system at his automotive plant. First the magneto ignition, then the whole engine, and finally the whole car is manufactured by the assembly lines Ford installs in 1913, reinventing not just the car-making process, but the whole industrial world.

1916 American physicist Paul Langevin, in England to help the British end the scourge of WWI German submarine attacks, hears of the piezoelectric effect, in which quartz crystals generate variable electrical responses to pressure. Langevin decides to see if a submersible quartz microphone might be fashioned to detect the sound of subs underwater. It works on the first try, and sonar (sound navigation and ranging) is born, saving countless lives during the world wars and beyond. Active sonar uses "pings" and reads their echoes to recognize silent objects—but it kills whales by disrupting their natural sonar systems.

1920 Gen. John Thompson invents the first machine gun light enough to be truly portable; uses force of each shot to load next round. Thompson gives it the name "submachine gun," but in the popular press it's the "Tommy gun." Prohibition gangsters made the gun infamous.

1920 New Jersey housewife Josephine Dickson keeps cutting herself, and keeps struggling with her cloth bandages. Inventor/engineer husband, Earle, makes up a batch of gauze patches on surgical tape. With the help of adhesive, Josephine applies the bandage without assistance. Earle Dickson works for Johnson & Johnson, and takes his new idea into work. The first commercial version is too big, but an improved marketing strategy packages the product in a variety of sizes. It is the Band-Aid.

1921 Charles Strite of Stillwater, Minnesota, is served burned toast once too often at his manufacturing-plant cafeteria. He invents the pop-up toaster by rigging a timer to a spring release and the heating coils. The Toastmaster, a refinement on Strite's patent, appears in 1926 and becomes ubiquitous after WWII.

Flash of Genius

Clarence Birdseye, who earned more than 300 patents during his busy life, investigated frozen food when he studied Eskimos and other Arctic populations as a naturalist for the U.S. government. He became interested in fur trading and, while on a 1916 research trip, noticed that fish frozen during the winter is much less mushy and much better tasting than that frozen during milder months. Natives confirm the phenomenon. After examining freezing and thawing foods under a microscope, he found that ice crystals ruin the character of most foods. But the dreaded crystals wouldn't form if the food was frozen quickly enough. He developed and patented a system of supercold disks surrounding a conveyor belt by which food is flash frozen. The technology establishes a $22 billion industry.

1924 Profound changes in eating habits arise when Clarence Birdseye of Brooklyn perfects a method of "flash freezing" food. He first markets fish and rabbit, then expands the line, but a public skeptical of the claim that frozen foods taste good isn't buying. Birdseye invests in hundreds of glass frozen-food cases to convince grocers and shoppers to try his products, and leases refrigerated railroad cars to distribute them nationwide. During WWII's rationing of "regular" food, Americans finally turn to frozen in large numbers and never go back.

1927 "You ain't heard nuthin' yet" are the first words spoken on the big screen, by Al Jolson, after Warner Brothers unveils its Vitaphone system, ushering in the era of talking pictures. New York premiere of *The Jazz Singer* shown here. FOX counters with the Movietone system, which evolves into the current standard. In Vitaphone, the sound is recorded and played back on discs in synch with the action on the screen. In Movietone, through the work of electronics pioneer Lee DeForest and others, cameras record sound right on the film; sound projectors can play it back.

1923–1927 The television set emerges according to a familiar pattern: An impractical first version leads to refinement stage of competing processes until finally the best process proves reliable and cost-efficient. Television first appears in 1925, when John Logie Baird shows a photoelectric system with spinning disks that successfully transmits a moving image. Soon, an all-electronic version is developed simultaneously by Russian-American Vladimir Zworkin and American Philo T. Farnsworth. Farnsworth's design wins out, forming the basis of the Television Age after WWII.

1926 Norwegian inventor Eric Rotheim discovers that liquid can be sealed into an aluminum container and then released under pressure, inventing the aerosol can. Not widely adopted until WWII when aerosols find wide use to apply pesticides. After the war, they enjoy a gigantic surge in popularity, but the widespread use of Freon (invented 1930 by Thomas Midgley) and other ozone-destroying chlorofluorocarbons brings controversy; United States bans CFCs in 1978; carbon dioxide, hydrochlorofluorocarbons substituted in spray cans.

1926 Robert Goddard invents an entirely new way to propel rockets: with liquid fuel. Launches the first one in Auburn, Massachusetts, in 1926, using liquid oxygen and gasoline. Dismissed as a crackpot and banned by the state fire marshal from further experiments after a 1929 explosion, he moves to Roswell, New Mexico. Encouraged by superstar aviator Charles Lindbergh, Goddard's liquid fuel leads to technologies that will terrorize London in WWII, take humankind to the Moon and beyond, threaten the Earth's existence, and enable satellite data transmission. Telephone, radio, television, digital data, photography, and cell phones are just a few of the innovations that Goddard's invention inspires.

Dr. Space

Robert Goddard earned his status as 1st to achieve rocket flight, and while he did most of his own work in isolation, he was far from the only one exploring the complex, dangerous, and promising technology of liquid-fueled rocketry. In fact, America's greatest leader in spaceflight development actually came from Germany. Amateur rocketry/astronomy clubs abounded in Germany and Russia before WWII; out of one of these clubs comes Wernher von Braun, a member of the Verein für Raumschiffahrt Society for Space Travel. It's shut down by the Nazis in 1932 and von Braun is impressed into the service of the Reich. Von Braun develops the horrific V-2 attack rocket but is arrested by the Gestapo in 1944 for preferring space research to the military; his usefulness wins his release. After the war, he becomes the lead developer of the American manned-space program, which eventually catches up with and surpasses its Russian rival. In 2003, readers of *Aviation Week and Space Technology* magazine name him the number-two figure in American aerospace history, second only to the Wright brothers.

1928 Otto Rohwedder makes the first sale of his bread slicer to a Missouri bakery. Consumers there think sliced bread is the greatest thing since . . . whatever the best invention was before that. But like so many inventors, Rohwedder fails to cash in on his idea and hard work. Demand from bakers for his slicer/wrapper is "fever-ish," but the Depression forces him to sell his invention and company and go to work for other manufacturers.

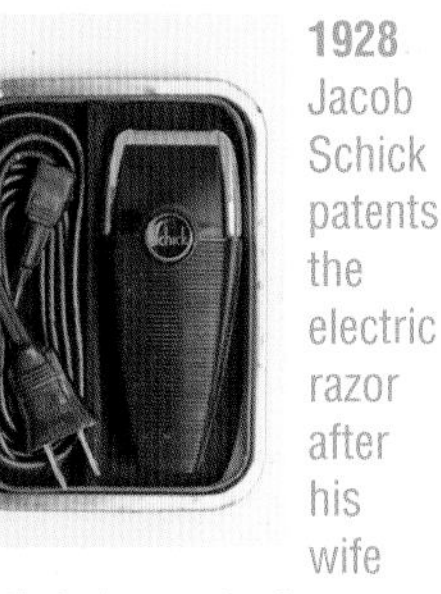

1928 Jacob Schick patents the electric razor after his wife mortgages their house to finance his research; his system of slotted vibrating blades is still the standard.

1929 Ernest Lawrence at University of California, Berkeley, invents and builds the cyclotron and uses it to accelerate atomic particles. The cyclotron's magnetic field guides electrically charged particles such as protons or electrons in near-circular paths while its electric field boosts them to speeds high enough so they penetrate the atomic nucleus (the atom's small dense core), probe its structure, form new radioactive isotopes useful in medicine and biology, and even create new chemical elements (plutonium, americium, einsteinium).

1932 Herbert Kalmus produces his 3rd version of the Technicolor process, and it's an enormous success. He invents the first Technicolor camera in 1916, using 2 colors (red and blue-green). This and the 2nd version are not too realistic, but the 3rd produces results audiences adore. Technicolor wanes because of its expense in the early part of the Depression, but *Snow White* in 1937 establishes its status as a moviemaking must.

1930

1934 Having vowed at age 15 to make a major mark on radio technology, 44-year-old Edwin Armstrong unveils a new radio tube he's rewired so it transmits by modulating wave width (frequency) instead of wave height (amplitude), for a huge increase in sound quality and clarity. FM radio is immensely popular, but Armstrong spends his life fighting patent battles; his wife wins the final case in 1967, long after his suicide.

1935 The British government assigns physicist Robert Watson-Watt to explore the lethal capabilities of radio. In proving that radio won't kill anyone, he makes a discovery that saves untold millions instead. He realizes blips in his recorded data are reflections of high-energy radio waves off solid objects. Proceeds to invent radar, which helps save Britain in WWII.

1936 In 1923, Spanish aeronautical designer Juan de la Cierva provides solution to a puzzle that had been vexing engineers and aviators since Leonardo's time: the physics of the helicopter rotor. His jointed rotator blades make possible the first practical helicopter model, the FW-61, designed by Heinrich Focke of Germany; Igor Sikorsky of Connecticut goes on to construct and then improve the helicopter as we know it today.

1937 Frank Whittle of England and Hans von Ohain of Germany invent the jet engine simultaneously and independently. Uses turbines to pull in air, which is combusted and shot out the back through more turbines, with enough force to achieve supersonic speed. Germany first to adopt the jet engine militarily, with Britain closely following, but jets don't really come into their own until late 1940s. Sound barrier broken October 14, 1947, by American Chuck Yeager.

1937 Having marveled as a kid at trapeze artists springing on their safety nets, a grown-up George Nissen spends 12 years making a trampoline that will sell well, is easy to set up, is fun, and suitable to home use. Promotes his 1938 version by mounting it on his car and touring the country.

Lost Chance

On the morning of December 7, 1941, radar shows a strong signal characteristic of a swarm of aircraft approaching the U.S. naval base at Pearl Harbor. When the report is passed on, poorly trained personnel assume the system is picking up American planes or is malfunctioning. A few hours later, much of the American naval fleet lies in ruins.

1937 Wallace Carothers and Julian Hill of the DuPont company invent a way to create an artificial silklike fabric when they learn how to stretch polymers, like polyester and polyimide, into long strands. The material gets stronger as it's stretched—stronger than steel by weight. Once they establish a method of spinning it, DuPont names it nylon, and a craze for synthetic fabrics sweeps America, founded on huge demand for nylon stockings.

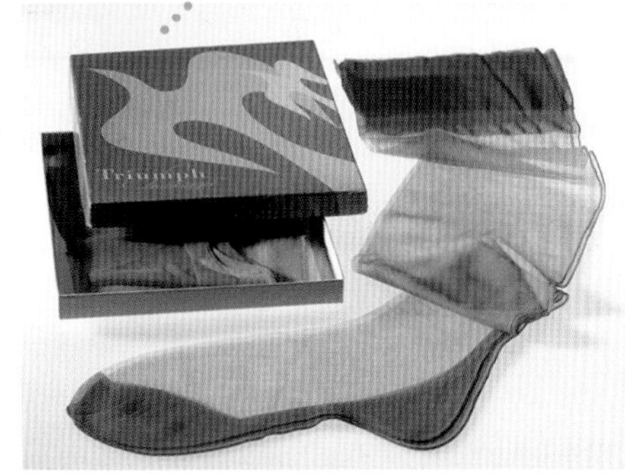

1938 The technique of pulling molten glass into fibers had been known for centuries; in 1836, Frenchman Ignace Dubus-Bonnel patented a method to weave hot glass on a loom. One hundred years later, fiberglass makes its first appearance after engineers at 2 glass corporations combine their efforts and the Owens-Corning Glass Company is founded. Besides insulation, fiberglass makes a very workable, light, strong building material when mixed with plastic.

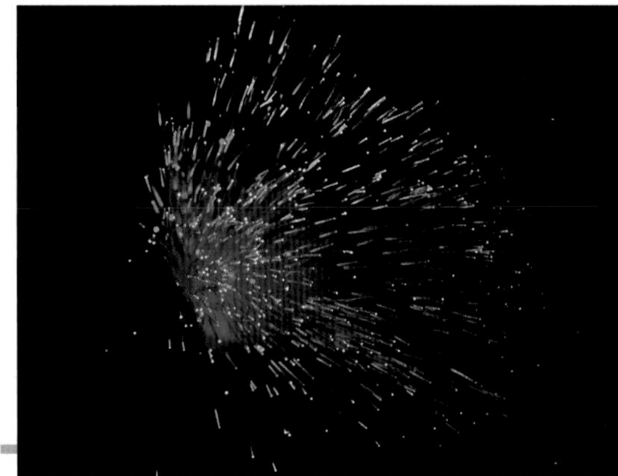

1938 Depression forces physicist Chester Carlson, shown here, center, to take a job copying documents by hand for an electrical firm. He researches copying technology in his kitchen; exploits light-sensitive electrical qualities of sulfur and, 5 years after starting work, successfully produces a copy of handwriting (the date and his location: "10-22-38, Astoria"). Haloid Co. buys license to develop his invention in 1944, changes its name to Xerox (*xerography* means "dry writing" in Greek).

1938 Brothers George and Ladislas Biro, a newsman and a chemist, join forces to create the ballpoint pen (still called *biros* in England). First version flops because ink delivery is poor, but brothers investigate capillary action and viscosity to create a model that proves popular. Franz Seech invents far superior ink in his kitchen and founds Paper Mate, 1949; Marcel Bich, Seech's former colleague at a failed ballpoint company, concentrates on bringing the cost down. His company, Bic, becomes world leader in ballpoint pens.

1940

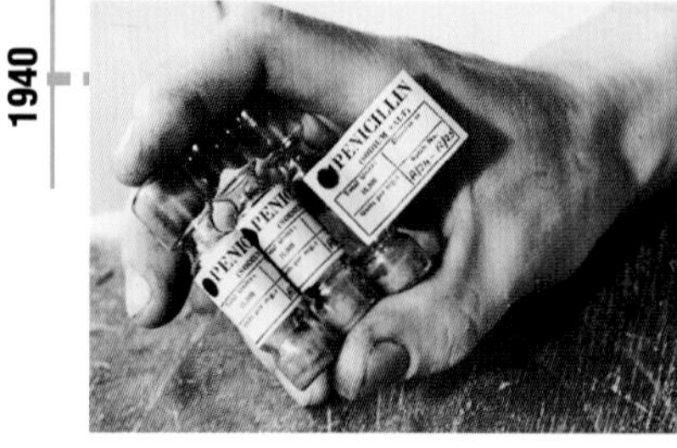

1940 Pathologist Howard Florey calls a fellow scientist with a simple message: "It's a miracle." Florey has injected 8 mice with deadly bacteria, and 4 with a mold containing penicillin. The 4 treated with penicillin are fine; the rest are dead. He works with scientists in the U.S. to find ways to extract the miracle drug, and a good process is developed in time to prevent thousands of combat-related infection deaths during WWII. Penicillin mold was identified and named in 1928 by Alexander Fleming of Scotland, but he didn't pursue his discovery until Florey and colleague Ernst Chain investigated and advanced his findings years later.

1942 Led by Italian physicist Enrico Fermi, an international team of scientists creates a sustained, controlled nuclear reaction in a squash court underneath the football stadium at the University of Chicago. Nuclear fission—splitting the core of the atom—works by smashing neutrons into uranium nuclei, which give off more neutrons. Work of top-secret "Manhattan Project" forms basis of atomic weapons and nuclear power; yields tremendous new insight about fundamental nature of matter and energy.

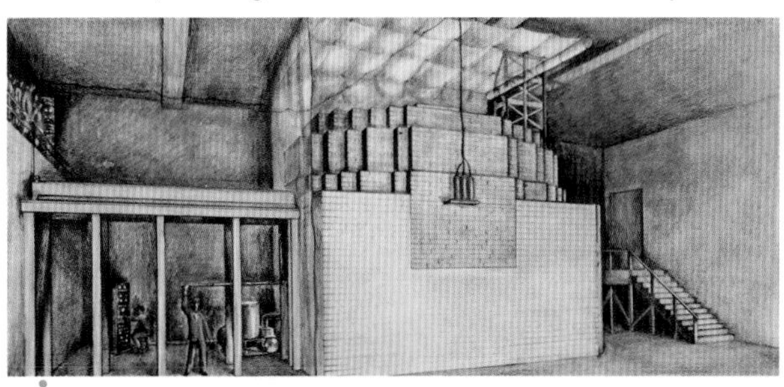

1943 French oceanographer Jacques Cousteau takes the science far beyond its historic level—literally and figuratively—when he invents the Aqua-Lung, a combination of compressed air tanks, tubing, straps, and a special pressure-regulating mouth valve. Enables extended stays underwater, opening the marine world to unprecedented exploration, by professionals and amateurs alike. Invention of scuba diving.

1944 Dutch physician Willem Koff constructs the first artificial kidney, a collection of wooden containers, laundry tubs, and tubes made from cellophane that can pump and clean blood of patients whose kidneys are failing. Within 20 years home dialysis machines such as this elegant model appear.

World War II

Wars spur invention, and World War II generated a great burst. Allied scientists beat the Germans to the atomic bomb, ensuring the Allies' victory over the devastating firepower of the Axis forces. But the technology war was waged on many other fronts, and the Allies didn't win every battle. But by 1944–45, Axis weapons that once appeared so cutting edge suddenly seemed slow, weak, and old-fashioned next to the Allies' advances. Weaponry and other technology leaped a generation in just 5 years. For each major improvement, like jets or computers, there were a dozen smaller other ones, each playing a role in winning the war and also advancing the state of the art in scores of industries and sciences. Some of the items that came of age or came into being as the world's engineers and industrialists devoted themselves to the cause:

Aircraft: jet engine, ballistic missiles, long-range bombers
Communications: microwave, radio phone, radar
Information science: circuit boards, digital computer
Medicine: "miracle" antibiotics, much-improved hygiene
Textiles and materials: plastics, including synthetic fabrics
Weapons: atomic bomb, submarine, aircraft carrier

1944 Howard Aiken's team turns on the Mark I at Harvard University' Lyman Lab, and the digital computer age begins. With a weight of 5 tons and measuring 8 x 50 feet, it performs any calculation accurately and automatically. First all-electronic computer (no mechanical switches), the ENIAC debuts at the University of Pennsylvania in 1946. The first commercially successful computer is UNIVAC, built by Remington Rand, Inc. in 1951.

1945

1945 Los Alamos National Lab scientists successfully detonate the first-ever atomic bomb at Alamogordo, New Mexico. They trigger this nuclear blast by compressing a sphere of plutonium with conventional explosives, proving it is feasible to make nuclear bombs using this man-made element. They also use a rare isotope of uranium to make the atomic bomb dropped on Hiroshima, Japan, on August 6. Physicists never test that bomb's design because they are sure it will work and had only enough uranium for one bomb.

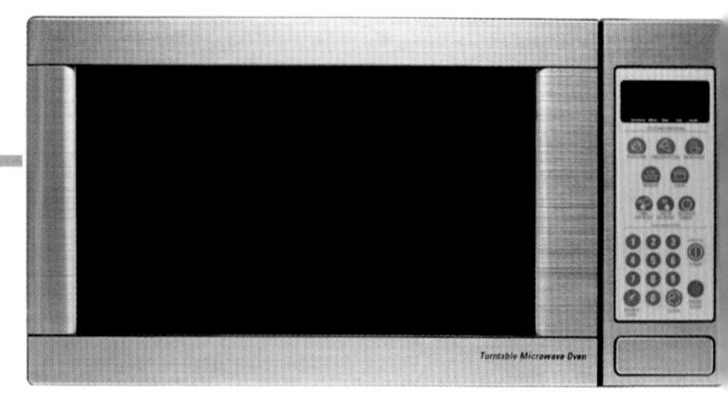

1946 Raytheon engineer Percy LeBaron Spencer is conducting research on his improved magnetron for radar when he realizes the chocolate bar in his pocket has melted. He suspects the microwaves (very high-energy radio waves) emitted by the magnetron are behind this strange development, and tests his theory with popcorn. Spencer invents not only microwave popcorn, but the microwave oven itself. Amana Radarange appearsin 1947.

Postwar Boom

The gains in technology and production capacity movitated by WWII greatly increased the standard of living when peace came again. In America, and eventually in a recovered Europe, improvements in production techniques made it possible to turn out a vastly expanded variety of products: cars, appliances, electronics, plastic goods of every kind. As demand spurs production, wages rise, which spurred demand, and this cycle created a standard of living far beyond anything the world had ever seen, and for a higher percentage of the population. But many African, South American, and Asian nations miss out, exploited for their raw materials and eventually, cheap labor, but stymied from acquiring the tools of production (power plants, communications grids) necessary to join the "First World." Legacy of colonial mistreatment persists into the 21st century.

1947 Physicists John Bardeen, William Shockley, and Walter Brittain determine, partly by accident, that certain semiconductors (germanium and silicon) have "holes" in their electrical resistance that mean they can be used like tiny electrical switches. Combining talents, the team uses their knowledge to create the "transfer resistor," renamed the transistor by a colleague at Bell Labs. They win the 1956 Nobel Prize for physics, and the transistor remakes the world. Basis of "solid state" electronics; end of the bulky, hot vacuum tube (shown).

Record Pace

The move from 78s to LPs was one of the many shifts in delivery of personal music that characterized the 20th century and continues in the 21st. The most profound change was the introduction of the phonograph by way of the gramophone, which brought art music to common people for the first time. The hand-cranked Victrola (1906) gave way to the electric record player; 78s introduced as standard in 1925. When CBS introduces LPs in 1948, archrival RCA countered with the "single" on 45 rpm records. After a few decades of stability, changes came quickly: Cassettes supplanted records, and then were supplanted themselves by compact discs, while boom boxes and personal players push aside record playes and home stereos; after the turn of the 21st century, Internet-based media and MP3 players form the basis of yet another tectonic shift. Innovations are the products of miniaturization, usually accomplished by teams of research-and-development engineers instead of the lone maverick typical of earlier periods.

1947 Mother Marion Donovan, up to her elbows in soaking cloth diapers and wet bed linens, is frustrated at recurring diaper rash that is caused by her children's rubber pants that lock in moisture. Sewing a layer of cotton to her shower curtain, she invents the "Boater," a disposable diaper, which debuts at Saks Fifth Avenue in 1948. About 19 billion are now sold annually in the U.S. alone.

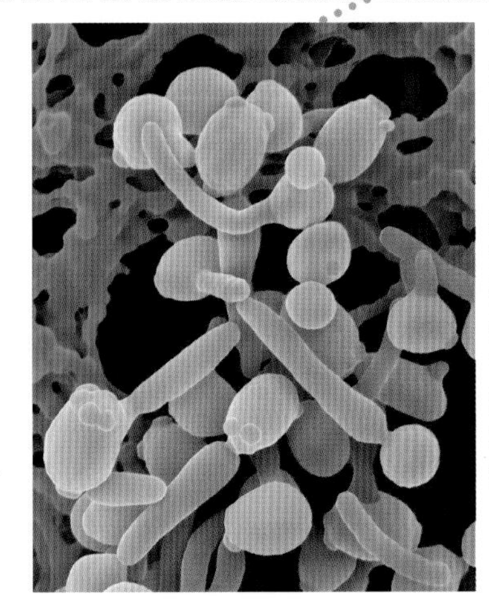

1948 While bacterial infections are being cured as never before, fungal diseases—from athlete's foot to fatal central-nervous system infections (and Candida albicans, or yeast, shown)—are actually on the rise until scientists Rachel Fuller Brown and Elizabeth Hazen invent Nystatin, the first effective antifungal. Assigned to the work by New York State government, the pair name their drug after their state and donate billions in patent proceeds to fund further research.

1948 Peter Carl Goldmark likes listening to music at parties, but dislikes that hosts have to keep flipping the 78 rpm records because so little music is contained on each one. After 3 years of work, Goldmark and CBS unveil the long-playing record, or LP, with microgrooves on vinyl, instead of the traditional shellac, for much better sound. The LP is the pinnacle of Goldmark's brilliant career, garnering 180 patents, most on behalf of CBS.

1948 Swiss electrical engineer Georges de Mestral goes hunting and comes back covered in burrs. He is inspired to create a fastening system that holds just as tenaciously, but can be easily released and reapplied. After eight years and many false starts (too much or too little hold, not durable enough, too expensive to make), de Mestral achieves his objective: a fastening system of tiny nylon hooks burrowing into soft nylon "velvet." For a name, de Mestral combines the French for "velvet" and "hook": velours + crochet = Velcro.

1948 Edwin Land produces the most successful in a series of inventions aligning crystalline substances to polarize light and filter out glare. Prompted by his daughter, impatient to see a photo of herself, Land develops a camera that makes prints almost immediately. Polaroid camera sets benchmark for home photography until introduction of digital cameras in late 20th century.

5 58242 66335 9

1949 Two business-school grad students hear their dean tell a grocer that an automated inventory and price-ringing system can't be built. Bernard Silver and Joseph Woodward take dots and dashes of Morse code and lengthen them into narrower and wider lines. Over 15 years, as lasers and computers catch up with their idea, they adapt it to the improving technology. All comes together in 1974, when bar code on a pack of gum is scanned in Troy, Ohio. Entire American retail industry now runs on universal product code (UPC) system.

1949 Harold Lyons of the U.S. National Bureau of Standards invents the molecular clock. He uses the oscillations of ammonia molecules to feed electromagnetic pulses to regulate the timepiece. Basis of the atomic clock (1955, cesium-based, shown here), which works on even more minute scale for even more accurate timekeeping. Current state of the art (always improving) is plus or minus 1 second every 15 million years.

1950 Banker Frank McNamara and his attorney, Ralph Schneider, create Diners Club, the first credit card, after McNamara is caught without his wallet at the end of a restaurant dinner. McNamara decides that people's ability to spend should be based on what they can afford, not the amount of cash they happen to be carrying. The revolution the men start is best described on the Diners Club Web site: "In 1950, a world without cash was inconceivable. Today, an economic universe without plastic is just as inconceivable."

1950

1950 Progesterone, the hormone that tells a woman's body she's pregnant and not to produce an egg that month, is synthesized through a process perfected by chemist Carl Djerassi, who calls his invention progestin. He spends another year making a version that's effective when taken by mouth. The birth-control pill, approved for U.S. use in 1960, ignites an enormous social reexamination as pregnancy truly becomes a choice and women are freed to establish a presence in male-dominated spheres of work and self-development.

The Pill

Catholics were forced to choose whether to defy their church's doctrine by taking "the Pill"; many did, leading to a crisis of American faith. Though the Pill elevated women's empowerment, it was, in no way, the first invention to change their status. From electric clothes dryers to frozen food, a host of technologies cut the time and effort needed to accomplish what was previously called "women's work," and this emancipated women to dare to travel other paths. Djerassi predicted correctly that his invention would have a profound impact on the genders and social and psychological relationships, and he said men would feel the impact more strongly than women as their traditional dominance was questioned and removed. Djerassi was in a position to hold forth on such questions; besides being a brilliant chemist and innovator (he also invented artificial antihistamines), he was an accomplished novelist, poet, playwright, and social observer.

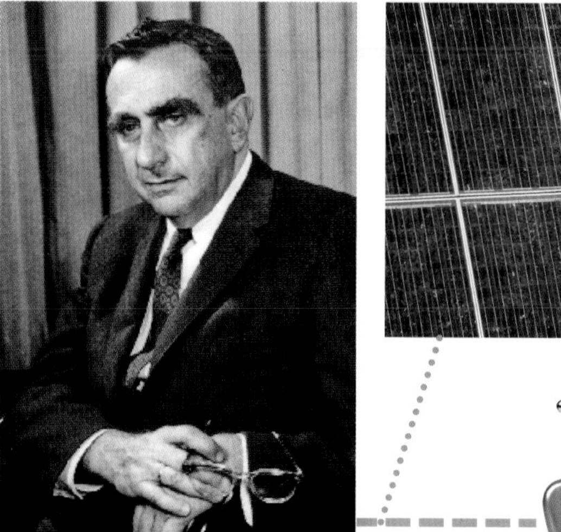

1950

1952 First hydrogen bomb tested on Enewetak Atoll in the South Pacific. "Inventors" are Edward Teller (shown) and Stanislaw Ulam, who lead a team in triggering nuclear fusion, the power source of the sun, which fuses hydrogen nuclei into helium. Major impetus for development of bomb was Soviet nuclear test in 1949, which stunned Americans. Soviets quickly test their own H-bomb, commencing the age of thermonuclear anxiety.

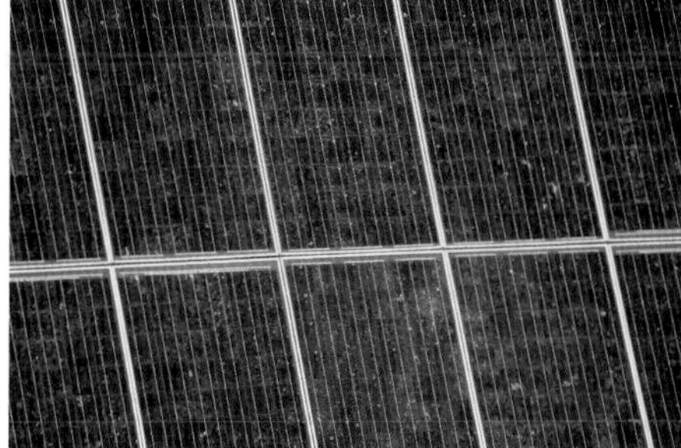

1954 Team of researchers at Bell Labs invents practical solar power about a century after Frenchman Edmond Becquerel discovers that sunlight can induce an electrical current (photovoltaic effect). Bell Labs makes 1st useful solar cell after finding that phosphorus and silicon combined properly can create a negatively charged zone on the cell, and silicon and boron a positively charged zone. In stacks, these wafers generate electricity when struck by sunlight. First solar power plant built in Hawaii in 1980.

1954 Texas Instruments President P. E. Haggerty commissions a transistor radio, which will demonstrate the power of the transistor and boost demand for TI's main product. The radio, built by the IDEA Corporation of Indianapolis, is marketed first by RCA (Radio Corporation of America); as RCA expects, it's a huge hit and introduces ordinary people worldwide to the everyday impact of the electronic age. As Japan's economy recovers from WWII, its corporations turn their attention to consumer electronics.

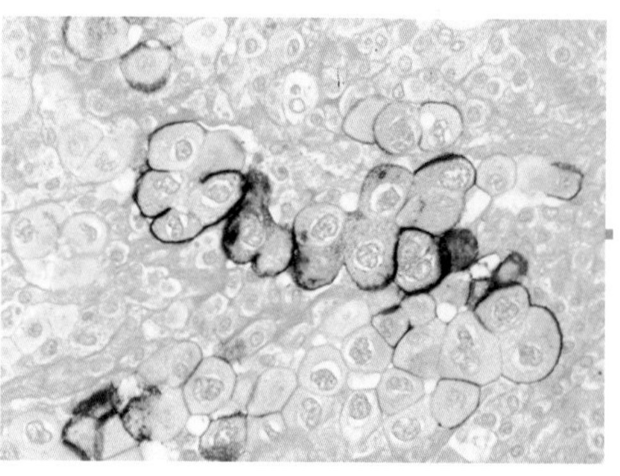

1955 Gertrude Elion of Burroughs Wellcome invents thioguanine and 6-MP, the 1st chemotherapy drugs. She constructs synthetic proteins that "convince" cancer cells to absorb them. But once inside, the proteins disrupt the biological processes of the cancer cells (shown, pancreatic cancer cells) without killing normal cells. Childhood leukemia greatly diminished.

1955 Lloyd Conover, a chemist with Pfizer Pharmaceuticals, invents tetracycline, the first "broad spectrum" antibiotic, i.e., a medicine that can wipe out lots of different disease-causing bacteria. Connover develops a method to artificially modify the structure of microbe-produced chemicals. Triggers wave of production of new antibiotics.

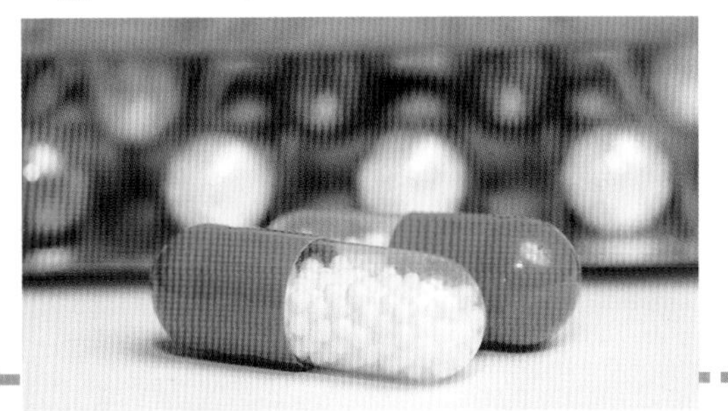

Radioimmunoassay

Pace of medical breakthroughs increased as biochemists mastered the realms of the molecule and atom. Prime example: Arguably the greatest medical invention of the 1950s was made by a nuclear physicist. Like Elion, Rosalyn Yalow got her chance at an advanced education, and the opportunity to do great work, mostly because WWII called so many male graduate students from the nation's universities. Yalow earns her Ph.D. from Illinois University in 1945, and studies the use of radioactive isotopes and human biochemistry. With Solomon Berson at the Veterans Administration hospital in the Bronx, New York, Yalow develops a way to measure hormone levels using radioactive isotopes (chemical elements with an unusual number of neutrons in their nuclei). The team soon realized they discovered an incredibly sensitive test for thousands of blood chemicals. Tests performed in a tube with a very small amount of blood; the effect on medical diagnosis is immense. In awarding the pair the 1977 Nobel Prize in physiology, Nobel Committee calls radioimmunoassay "more important than the discovery of X-rays."

1955 The scourge of polio, which leaves Franklin D. Roosevelt in a wheelchair and cripples tens of thousands of children a year, meets its match when Dr. Jonas Salk (shown) creates a workable vaccine. He becomes an international hero, and schoolchildren are immunized for polio at a frantic rate. Soon afterward, Dr. Albert Sabin invents a superior oral vaccine for polio; the Salk/Sabin feud joins the list of all-time great technological disputes.

1955 Englishman Christopher Cockerell builds the first boat that rides on a cushion of air, making the hovercraft possible. He builds a demonstration model out of a vacuum cleaner motor and cat food cans. Passenger service starts 1959; maiden voyage shown here.

1956 Ampex hires Charles Ginsburg and Ray Dolby to tackle a problem that has stumped the best electrical engineers for two decades: how to capture a magnetic print of a video signal so it can be stored on tape. Solution: a rotating "head" records the signal on a horizontally moving tape, greatly increasing the storage space on each inch of tape. Up until now, television news crews shoot film that is developed at the TV studio and rolled during a newscast; by the 1970s, new video technology replaces this cumbersome process.

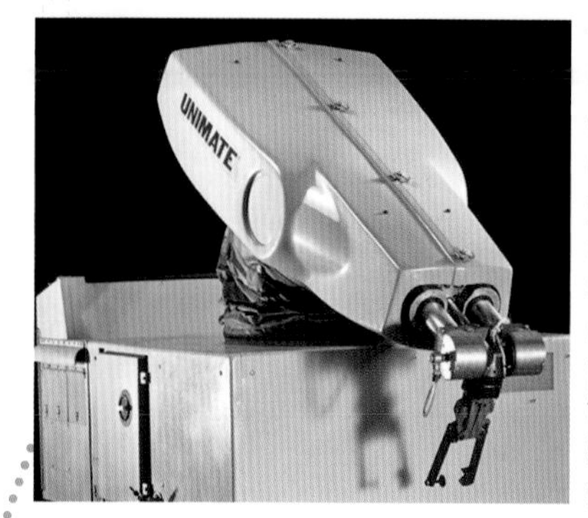

1956–1960 Automation experts Joe Engleberger and George Devol meet to discuss *I, Robot* and other works on robotics by Isaac Asimov. Over several years they build a robotic arm that can detect if the motion it's made has accomplished its goal, and if not, can make adjustments so the task is completed correctly. General Motors buys the Unimate in 1960 and installs it on its assembly line, where it performs perfectly and displaces human workers.

1957 Soviet rocket scientists Valentin Glushko and Sergey Korolyov lead a team that launches Sputnik, the first artificial satellite (shown), into space October 4, initiating the Space Age. The 184-pound sphere has 4 antennae that transmit data about the atmosphere, along with radio beeps on 2 frequencies that ham radio operators worldwide can capture on their sets. It orbits the earth once every 95 minutes until January 4, 1958.

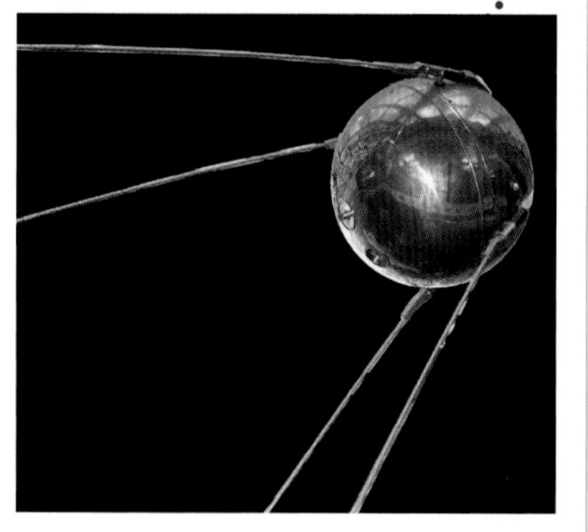

Space Race/Cold War

Launch of the Sputnik was a huge triumph for the Soviet Union and its scientists, locked in a battle for world supremacy with the U.S. and other Western powers. The U.S. assumed it would be the first into space. The radio beeps from Sputnik, picked up and broadcast nationwide, sounded to a lot of Americans like a menacing taunt. Much was also made of the fact that the Soviets' satellite was far heavier and robust than the one U.S. designers were preparing. The shock of Sputnik sparked a massive new interest and investment in science and technology education. The wonders of the Space Age—tiny communications equipment, lightweight plastics, computer technology—intermingled with a fearful dark side: The same rockets that could explore the heavens could also carry nuclear weapons across the oceans in a matter of minutes, and the space and arms races were tied together. This coexistence of hope and fear, as communism in the East battles capitalism in the West, defined a generation. A whole new body of popular art culture devoted to secret agents and space warriors reflected the psychology of the times.

1957 Joseph-Armand Bombardier of Valcourt, Quebec, achieves his lifelong goal of making a fast, reliable, and marketable snowmobile by attaching rubber tracks to a set of sprocketed wheels and placing short skis in front. The Ski-Dog—named to replace the dogsled—is born. Redesigned in 1959 and renamed Ski-Doo.

Light Show

The public imagination identified lasers with death rays and thick steel beams dropping apart like butter. Laser's true utility came from its precision, which stemmed from the identicalness of the light waves it generates. That said, most governments have been and will continue to be extremely interested in the laser's potential as a weapon. Light sabers may be in the (not too distant) future; in the meantime, the number of applications for lasers is already astonishing: price scanners, CDs and DVDs, computer printers, medical uses such as Lasik surgery, materials processing, alarm and automation sensors, and telecommunications; and the list is growing fast.

1957 Belgian Basil Hirschowitz, a medical researcher hired by the University of Michigan, decides to try to improve exploratory surgery, which is almost useless because there's no method to illuminate internal organs, tissues, and structures. He and graduate student Larry Curtiss reinvent fiber-optic cables out of silicon dioxide so they won't break. Makes endoscopic surgery practical, and telecommunications industry adopts their design for optic cable.

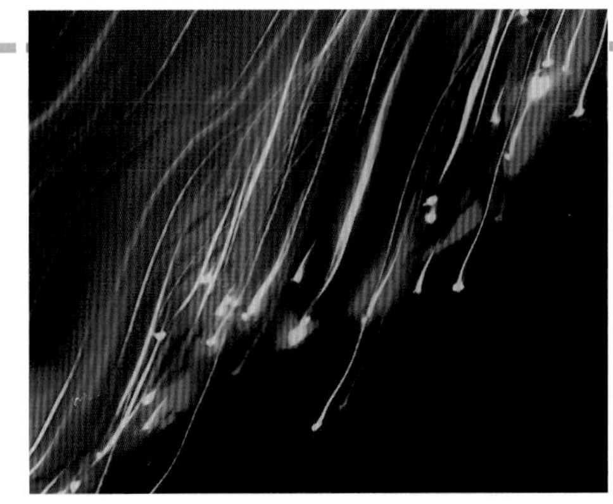

1957–1960 Laser created by physicist Gordon Gould (1957); first demonstrated by Theodore Maiman in 1960. "Laser" is not the beam, but the machine that creates it. Atoms are forced into an excited state. When they return to normal, they shoot light waves that excite other atoms and generate identical waves; these identical waves don't scatter like ordinary light does. Focused out of the laser, beams of such light can easily cut diamonds and steel, or measure to accuracy of 1/100 of a human hair.

1958 Two U.S. researchers realize the amazing potential of a chemical discovered in 1942, and exploit it to invent superglue. The chemical is cyanocrylate, and it turns into plastic on contact with even a tiny amount of water. Only one drop of glue can hold hundreds of pounds and bond fingers instantly. One of the researchers is Harry Coover and the other has the perfect name for an inventor of glue: Fred Joyner.

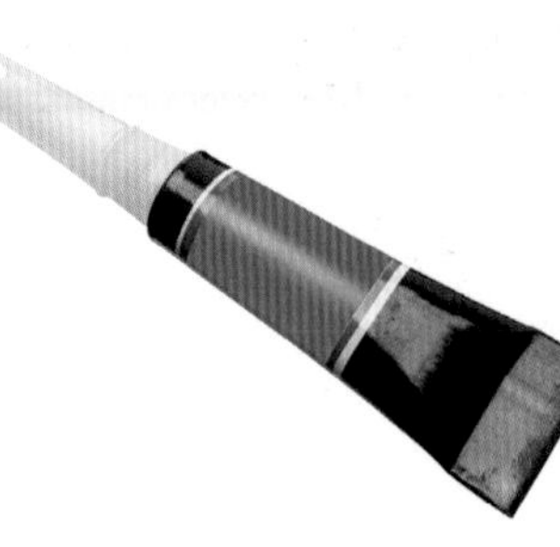

1958 Electrical engineer Jack Kilby of Texas Instruments works almost alone in the TI lab during summer vacation season to create an integrated circuit, a single piece or chip of silicon that can carry all the tiny connections needed to make miniaturized electronics possible. He succeeds, but immediately is enmeshed in one of history's great patent/priority battles, with Robert Noyce of Fairchild Semiconductor. Noyce wins the patent; Kilby wins most of the recognition. Modern electronics is born.

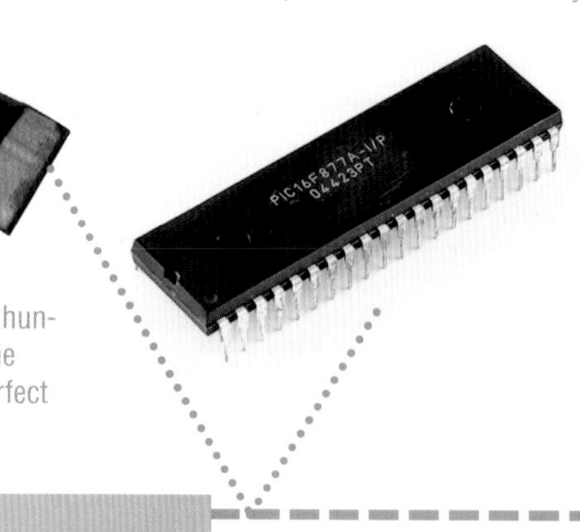

1958 Swedish engineer Nils Bohlin invents the lap-and-shoulder seat belt familiar today, and it appears in cars made by his employer, Volvo.

1958 Evolution of the transistor makes cardiac pacemaker possible. Wilson Greatbach, recipient of 150 patents, uses transistor to implement his idea: deliver regular shocks to the sinus node of the heart, which receives the "beat" impulse from the brain. In heart blockage, sinus node can't get the electrical signal from the brain, but Greatbach's device provides a substitute. His first working model is ready 2 weeks after acquiring the right transistors; first implanted 1960; hundreds of thousands of lives saved since.

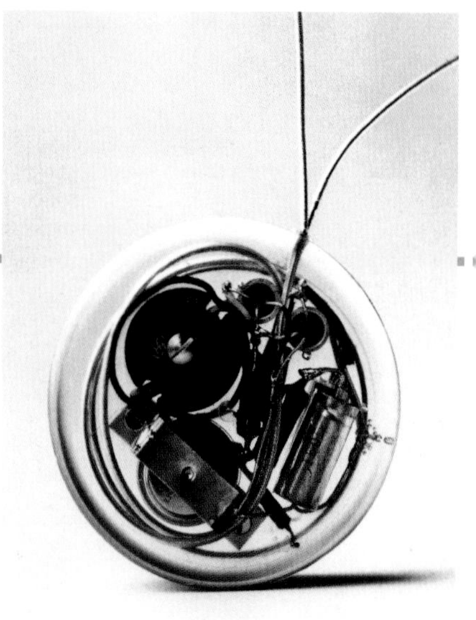

Bell Labs

Bell Labs and "the future" meant the same thing in the popular mind. Established in 1925 by Walter Gifford, president of Alexander Graham Bell's company AT&T, the labs defined research-and-development excellence for 75 years, especially during the miracle-technology years of the 1950s and '60s. Transistors, lasers, solar cells, and satellites were developed in Bell's Murray Hill, New Jersey, headquarters. Staffers have won 6 Nobel prizes, and inventions from the fax machine to the calculator to Voiceover-Internet-protocol have streamed forth. If anyone could take credit for inventing the Internet it's the employees of Bell Labs, who developed its data-transmission backbone.

1961 Soviets beat Americans again; Sergey Korolyov and his team of engineers launch Yuri Gagarin into Earth orbit on the *Vostok 1*; he travels once around the planet in 108 minutes. Invention of manned space flight; spurs U.S. President Kennedy to set goal of American on the moon before 1970.

1960

1960 U.S. engineer John Pierce of Bell Labs leads a team in bouncing a voice-carrying radio signal off Echo I, a highly reflective balloon. It performs poorly but demonstrates feasibility of communications satellites. The Pierce team deploys Telstar I, launched by NASA on behalf of AT&T, in 1962 to relay voice and data of every kind, i.e., satellite radio, around the world.

1962 Diodes are devices that get current flowing in one direction; their "on-off" property is the basis of electronics. The most familiar form in ordinary modern life is the LED: light-emitting diode. After studying with John Bardeen, one of the inventors of the transistor, Nick Holonyak of General Electric rigs a transistor to make the current excite a phosphorescent material (gallium arsenide most commonly used today). High-resolution rotating and holographic displays the cutting edge of LED technology. Tiny, eyeball-size monitors likely to become common soon.

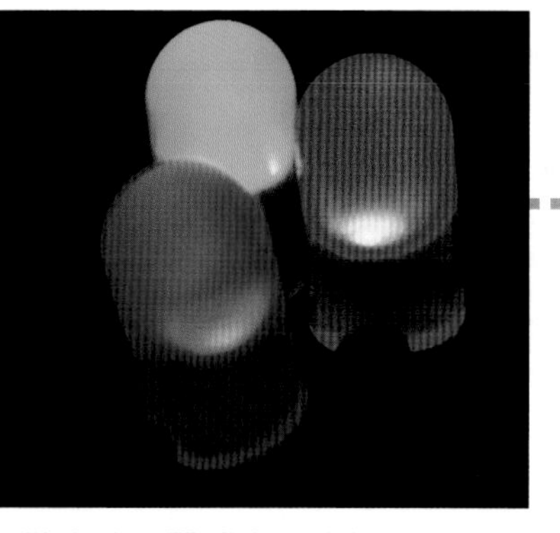

1967 John Shepherd-Barron of De La Rue Instruments realizes his vision when Barclays Bank of London dispenses a 10-pound note from a machine on June 27. First ATM activated by carbon-treated slips obtained in advance from the regular bank; Don Wetzel, head of product planning at Docutel in the U.S., is inspired to invent card-operated version while standing in a bank line in Dallas.

1968 As he did with the transistor, Texas Instruments president Patrick Haggerty thinks of a way to show off the capability of integrated circuits. He sets their inventor, Jack Kilby, to the task of inventing a calculator that can be carried in a pocket. Kilby loves meeting this type of challenge. Weighing 2-1/2 pounds, the Pocketronic Printing Calculator, released in 1972, is a huge success. Soon thereafter, LED technology catches up with the pocket calculator idea, and the slide rule's days are numbered.

1970 The Hamilton Watch company builds first digital watch based on LED and integrated-circuit technology, ready for commercial sale in 1972. Digital watches are frequently handed out free as promotions today; in 1972, the Pulsar cost more than $1,000. LED watches are soon displaced by those using another technology, invented in 1968: liquid-crystal displays, which use current to arrange crystals that line up to form characters with the use of treated, polarized glass, expending almost no power.

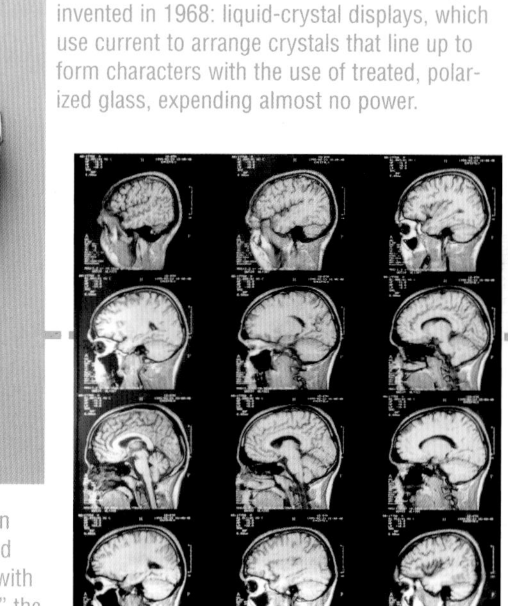

1977 Scientists begin extracting very detailed images of molecules with "magnetic resonance," the distinguishing behavior of atomic nuclei in dense magnetic fields. In 1970, Dr. Raymond Damadian realizes the technique might be used to examine living tissue. After a long series of experiments and refinements, he and his team build the first fully functional diagnostic MRI machine, call it the Indomitable because so many problems are overcome to build it. Incredibly useful, and therefore used abundantly, in medicine. Brain scans seen here.

1977 Ken Olsen, who capitalizes on improvements in integrated circuits to invent the mid-sized "minicomputer" in 1965, had scoffed at the idea of a computer that could fit on a desk. But Steve Wozniak of California sees the future and makes it happen. Working with microchips in his garage in Palo Alto, California, he and Steve Jobs develop a prototype that proves the personal computer could work. Their next model, the Apple II, inaugurates a new era in human living, turning the computer into an essential tool for everyday work and play.

Alone at Last

A compact disc consisted of a plastic disc coated with aluminum or other metal. A recording laser embossed peaks and valleys in the metal, and those were then read by the laser in the player, and converted into sound waves through circuit boards and speakers. Sony's Walkman personal cassette player (launched 1979) quickly adapted for CD with the Discman; the iPod in 2001 brought all-digital media to the same platform. The personal headphones on these devices institute an entirely new mode for experiencing music and other sound: privately, silently, in isolation from the world while moving through it publicly. It is a new social (antisocial?) as well as artistic and technological phenomenon.

1979 Radio phones were in use during WWII, and the car phone appears at the end of the 1970s, but the bulkiness of receivers and the limited number of transmitters keep it from becoming a tool for every-one until the FCC divides urban areas into cells with lots of microwave frequencies. Phones are quickly developed that constantly retune the user from one cell to another as he or she travels. Bell Labs and AT&T lead the way in this new, and now ubiquitous, cellular phone technology. Most calls now sent digitally (streams of zeros and ones representing "off" and "on" electronic states) instead of as analog signals like waves or currents.

1980 With a computer monitor about to appear on every desk, Arthur Fry comes up with something to put on every computer monitor: the Post-it note. His friend at 3M, Spencer Silver, has made an extraordinary weak adhesive in the process of searching for an extraordinarily strong one. Fry thinks it might be just the thing to hold bookmarks in place without damaging the book. Eventually sold as notepads, "press-and-peels" become an invention people can't do without.

1982 Compact disc's digital method of storing sound was first proposed about 1800 by Jean Joseph Fourier, who theorized sound waves could be represented as on-off patterns taken from samples of the wave. The full implementation of this old idea came with the maturation of the laser and the development of "information theory," notably at Bell Labs and MIT. Magnavox introduces videodisc, or laserdisc, in 1978, and from there it is not long before the CD is ready for the mass market.

1990 Tim Berners-Lee, seeking to better coordinate information sharing over the fledgling Internet, writes a standard for linking documents and names it the World Wide Web. Exploits hypertext, invented in 1965 by cyberarchitect Ted Nelson to enhance document software by making cross-references clickable. In 1993, Marc Andreessen and Eric Bina invent the first browser, Mosaic, which displays pictures along with text, and the fundamentals of a commercial, social, and communications revolution are in place.

1987 Dr. Ray W. Fuller leads a team of pharmacologists at Eli Lilly through a long process to invent a chemical that will block the uptake of serotonin, a primary "feel-good" neurotransmitter. They succeed by creating fluoxetine, which wins U.S. Food and Drug Administration approval as the prescription antidepressant Prozac. Prozac engenders a fundamental shift in popular attitudes toward managing psychological difficulties, particularly depression, with medication.

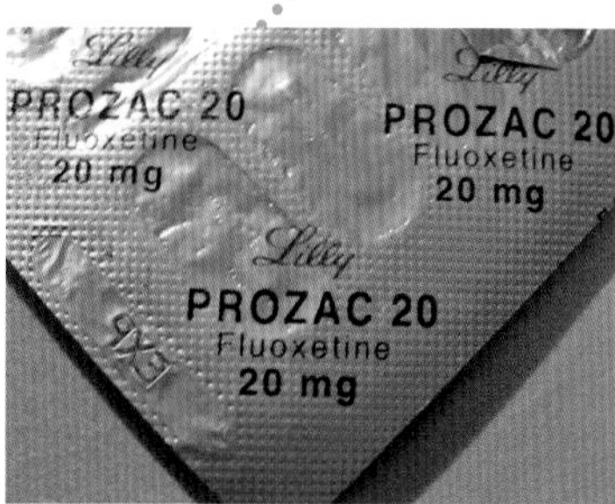

1994 Jeff Bezos sets up Cadabra.com, which he envisions as a way for people to select books and pay for them through their computers. Launches the Web site the following year with a new name: Amazon.com. The "Internet bubble" wave of overinvestment in the new commerce model follows, and skeptics believe Amazon.com will never succeed. It finally turns a profit in 2003, and the era of profitable e-commerce begins.

1996 After 277 tries, a team at Scotland's Roslin Institute succeed in inventing the cloned animal. They place a stem cell in the egg of a sheep, then use the egg to impregnate a normal ewe. When Lamb 6LL3 is born, they name her Dolly. One main goal of cloning: to replicate animals that are genetically modified to produce beneficial medical substances. In 2001, the first member of an endangered species, the oxlike gaur, is cloned from a dead adult.

2000

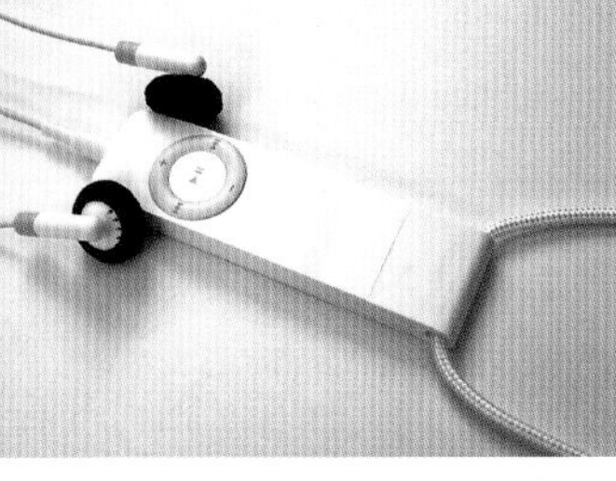

2001 "One thousand songs in your pocket" is how Steve Jobs introduces the iPod, invented by Jon Rubenstein, Anthony Fadell, and Stan Ng at Jobs's behest. The iPod, or the MP3 player, isn't actually a new technology—just a brilliant combination of existing ones—and, combined with a burgeoning Internet infrastructure for content (iTunes and podcasts leading the way), the iPod becomes the benchmark of cool technology in an age brimming with status-conferring hip gadgets.

2007 Hundreds of useful, innovative Web sites show how the Internet becomes the most accessible platform for invention ever devised. Wikipedia, a collaborative online encyclopedia, founded by Jimmy Wales, Larry Sanger, 1999; MySpace by Brad Greenspan, 2003; YouTube by Chad Hurley, Stephen Chen, 2005. Google, eBay reshape research and world commerce, and MySpace, MeetUp do the same for social networking. YouTube, a video-sharing Web site, named 2006 invention of the year by *TIME*, purchased by Google in 2006 for $1.65 billion. The possibilities for information and knowledge sharing rival the Gutenberg revolution of the 1600s.

The Future

At the end of any timeline, the eternal question arises: What's next? That's unknowable, of course, but current trends are heading in certain directions, making the future at least somewhat predictable. It's reasonable to expect development of the following:

- Magnetic-levitation (electromagnetic suspension) trains running coast to coast in America
- Electricity generated by nuclear fusion, with inexaustable, free hydrogen as fuel
- Self-cleaning glass by 2015
- Disposable computers
- Machines operated by the electromagnetism of brain waves
- Replacement body parts grown in the lab for specific patients
- Robotic domestic and commercial assistants
- Cloned humans

The process of technology improvement, always accelerating, seems certain to arrive someday at a pinnacle even more fantastic and disturbing than mass-produced human clones. Eventually, inventors will produce a device that produces inventions. If the past is any guide, this strange moment will arrive sooner than we can imagine today.

Photo Credits

Cover: Lit Lightbulb: Matthais Kulka/Adobe Stock Photos, Masterfile

Prehistoric Flint Weapons: The Granger Collection, New York; **Tanzania: Lion Dance**: The Granger Collection, New York; **Pile of Red Ochre Powder**: Erick Nguyen/123rf; **Greek Warship:** The Granger Collection, New York; **Flintstone Structure**: Oliver Lantzendörffer/123rf; **Assyrian: Musicians:** The Granger Collection, New York; **Wooden Hut in Asia**: Aman Khan/123rf; **Pickax from the Ancient Salt Mine at Hallstatt, 9th–10th BC**: Erich Lessing/Art Resource, NY; **Ishango Bone**: Science Museum of Brussels/Wikimedia Commons; **Aztec Atlatl**: Werner Forman/Art Resource, NY; **Cup to Measure Grain, 2200–1900 BC**: Erich Lessing/Art Resource, NY; **Magdalenian Harpoons**: The Granger Collection, New York; **Cave Painting of Bowman and Deer**: Art Resource, NY; **Neolithic Sickles, c. 3000 BC**: The Granger Collection, New York; **Hand–shaped and Hand–dyed Reed Splints with Twill Weave Construction**: Smithsonian American Art Museum, Washington, DC/Art Resource, NY; **Spear-head, from Amorgos**: Erich Lessing/Art Resource, NY; **Celtic Crosses, Necropolis, Glasgow**: Stephen Finn/123rf; **Sickle Blades, Terra–cotta, Obeid Epoch (4000–3500 BC)**: Réunion des Musées Nationaux/Art Resource, NY; **Mirror with Handle Depicting Hathor**: Scala/Art Resource, NY; **Babylonia: World Map**: The Granger Collection, New York; **Flax Bed:** Vaida Petreikiene/123rf; **Mesopotamia: Drinking Beer**: The Granger Collection, New York; **Boats and Barges on the Grand Canal at Suzhou**: Werner Forman/Art Resource, NY; **Egypt: Shadoof Irrigation**: The Granger Collection, New York; **Team of Plowing Oxen**: Scala/Art Resource, NY; **Merchant Carrying a Pair of Scales for Weighing Metals**: Erich Lessing/Art Resource, NY; **A Pair of Earrings with Cloisonne Ornament, Gold**: Erich Lessing/Art Resource, NY; **Bronze Age**: The Granger Collection, New York; **Prehistory: Gold Collar**: The Granger Collection, New York; **The Royal Game of Ur, from Ur, Southern Iraq, c. 2600–2400 BC**: HIP/Art Resource, NY; **Stack of Plywood**: © Thinkstock/CORBIS; **Clay Tablet with an Inscription with the Name King Samshi-Addad, Mentioning the Building of a Temple in Deir es Zor**: Erich Lessing/Art Resource, NY; **Assyrian Soldiers with Ox–cart Leading Elamite Prisoners of War to Exile**: Scala/Art Resource, NY; **Egypt: Scribe's Tool**: The Granger Collection, New York; **Group of Bronze Tools for Woodworking, from Thebes, Egypt, New Kingdom, c. 1300 BC**: HIP/Art Resource, NY; **A Model of a Sailing Boat with the Pilot in the Bow and the Owner Resting Under a Canopy**: Werner Forman/Art Resource, NY; **Pottery Manufacture**: The Granger Collection, New York; **The Road to Ancient City of Dan, Israel**: Erich Lessing/Art Resource, NY; **China: Silk Manufacture**: The Granger Collection, New York; **Roman Wall Construction**: Erich Lessing/Art Resource, NY; **Natural Colored Soaps with Nail Brush and Pumice Stone**: Gina Smith/123rf; **Temple of Amun, Hypostyle Hall**: Vanni/Art Resource, NY; **Two Rams in Dark Blue Overlay and the Inscription "Reptah" (Good Luck)**: Erich Lessing/Art Resource, NY; **Iraq: Sumerian Temple**: The Granger Collection, New York; **Historical Chain from a Castle**: Holger Wulschlaeger/123rf; **Chinese "Love" Character and Calligraphy Brushes on Paper**: © Gareth Brown/CORBIS; **Electrum Nugget, c. 1100 BC**: The Granger Collection, New York; **Top View of Colorful Chinese Silk Parasols**: James Steidl/123rf; **Gymnasia of P. Vedius Antonius**: Erich Lessing/Art Resource, NY; **Candle fire over Black**: Anatoly Tiplyashin/123rf; **Iron Age Pulley (c. 900 BC), Israelite, from Gezer**: Erich Lessing/Art Resource, NY;

Sailing Ship and Lighthouse from Cylindric Ara Rostrata (Dedicated to the Calm at Sea), Roman: Erich Lessing/Art Resource, NY; **Siren Painter (5th BC)**: Erich Lessing/Art Resource, NY; **Rubber Sap**: 123rf; **Phoenecian Cuneiform**: The Granger Collection, New York; **Gold Model Chariot from the Oxus Treasure, Achaemenid Persian, from the Region of Takht-I Kuwad, Tadjikistan, 5th–4th Century BC**: Erich Lessing/Art Resource, NY; **Afghani Style Sandal for Males**: Aman Khan/123rf; **Roman Coin: Caligula**: The Granger Collection, New York; **Sundial Made of Stone**: © Bettmann/CORBIS; **Alexandria: Library**: The Granger Collection, New York; **Greek Coin, 6th Century BC**: The Granger Collection, New York; **Set of Surgical Instruments**: Milos Luzanin/123rf; **Engraving of Euclid, c. 300 BC, Greek Mathematician**: Image Select/Art Resource, NY; **Antique Iron Boat Anchor Free Standing**: Roy Hulsbergen/123rf; **General View, Via Appia**: SEF/Art Resource, NY; **Abacus, Chinese Retro Calendar**: Carl Stone/123rf; **Roman Warfare:** The Granger Collection, New York; **Ancient Roman Water Pipes**: The Granger Collection, New York; **In a Martini Cup**: Yanik Chauvin/123rf; **Venetian Crossbow with Gaffle**: Erich Lessing/Art Resource, NY; **Chain Links**: Ira Struebel/123rf; **Bust of Archimedes**: Alinari/Art Resource, NY; **Archimedes' Screw for Raising Water from One Level to Another, 1815, Invented by Archimedes (c. 287–212 BC)**: HIP/Art Resource, NY; **World Map: Eratosthenes**: The Granger Collection, New York; **Canterbury Codex Aureus**: The Granger Collection, New York; **Compass: Chinese**: The Granger Collection, New York; **Calculator**: Pablo Eder/123rf; **William Henry Toms (18th C.)**: **The Chelsea Water–works Engine**: The LuEsther T. Mertz Library, NYBG/Art Resource, NY; **Leonardo: Odometer**: The Granger Collection, New York; **Floor Heating in the Caldarium of the Various Thermae–eleuthera Agora**: Erich Lessing/Art Resource, NY; **Glass Bottles, Eastern Mediterranean, Early 1st Century BC**: The Jewish Museum/Art Resource; NY; **Props and Rudders of Fishing Boats**: Adrian Hughes/123rf; **China: Paper Manufacture**: The Granger Collection, New York; **Harvest Season, c. 1466**: The Granger Collection, New York; **Pont du Gard, Ancient Roman Aqueduct and Bridge Spanning the Gardon River near Nime, France**: Vanni/Art Resource, NY; **Astrolabe (Sun Clock)**: Alexey Biryukov/123rf; **Loading a Hunting Rifle with Gunpowder**: © Michael Freeman/CORBIS; **Old Grinding Mill for Coffee**: Ariusz Nawrocki/123rf; **Rembrandt: Drapers Guild**: The Granger Collection, New York; **Colorful Sailboat**: Stephen Coburn/123rf; **Chinese Currency**: The Granger Collection, New York; **Huddersfield Narrow Canal, Northern England**: Tom Curtis/123rf; **Fork in Ivory and Steel**: Réunion des Musées Nationaux/Art Resource, NY; **Old Wooden Printing Type**: Ron Sumners/123rf; **Old Frame and Mirror**: Nicolas Nadjar/123rf; **Weapons: Mangonel**: The Granger Collection, New York; **The Apartment in Kettenbrueckengasse, Vienna, where Franz Schubert Died in 1828**: Erich Lessing/Art Resource, NY; **Firing a Blunderbuss, from Illuminated Manuscript, 15th Century**: Erich Lessing/Art Resource, NY; **Siege of Medieval Town**: The Granger Collection, New York; **Sandglass, 16th Century**: The Granger Collection, New York; **France: Iron Furnace**: The Granger Collection, New York; **Spanish Ship, 1496**: The Granger Collection, New York; **The Printing Press Invented by Johannes Gutenberg Between 1397 and 1400**: Erich Lessing/Art Resource, NY; **Anemometer, 20th Century**: The Granger Collection, New York; **Knots in Rope over Water**: © Skip Nall/CORBIS; **Blue Toothbrush**: 123rf; **Old Vintage Clocks, Watches**: 123rf; **Etcher and Assistant**: The Granger Collection, New York; **Naval Ship: Great Harry**: The Granger

Collection, New York; **Musketeer**: The Granger Collection, New York; **Morton Inhaler**: The Granger Collection, New York; **Set of Wooden Pencils**: Alexander Ryabchun/123rf; **Globe: Mercator, 1541**: The Granger Collection, New York; **World Map: Ortelius**: The Granger Collection, New York; **Monitor vs. Merrimack**: The Granger Collection, New York; **Surveying Students Practicing**: © Chris Sattlberger/CORBIS; **Standing Clock, 1589**: The Granger Collection, New York; **Pope Gregory XIII (1502–1585)**: The Granger Collection, New York; **Marcus Tullius Cicero**: The Granger Collection, New York; **Textile Manufacture, 1770**: The Granger Collection, New York; **Galileo's Microscope**: The Granger Collection, New York; **Galileo Galilei's Thermoscope**: Scala/Art Resource, NY; **Sir Francis Bacon (1561–1626)**: The Granger Collection, New York; **News Vendor, 1631**: The Granger Collection, New York; **Telescope, Triangle, Magnet Compass, and Pendulum Clock Belonging to Galileo Galilei (1564–1642)**: Erich Lessing/Art Resource, NY; **Napiers Bones, a Compass, a Square, and Some of Isaac Newton's Manuscripts**: Erich Lessing/Art Resource, NY; **Van Drebbel Submarine**: Royal Navy Submarine Museum, Hampshire, England; **"La Pascaline," Calculating Machine Constructed by Blaise Pascal, and Technical Drawings, 17th AD**: Erich Lessing/Art Resource, NY; **Calipers Measurement Tool**: Hans Slegers/123rf; **Slide Rule**: The Granger Collection, New York; **Evangelista Torricelli**: The Granger Collection, New York; **Otto von Guericke**: The Granger Collection, New York; **Dagger and Plug Bayonet**: Victoria & Albert Museum, London/Art Resource, NY; **Christian Huygens (1629–1695)**: The Granger Collection, New York; **Clock Main Spring**: © Comstock/CORBIS; **Royal Society, 1667**: The Granger Collection, New York; **Pencil on a Calculus Problem:** Uyen Le/123rf; **Denis Papin (1647–1712)**: The Granger Collection, New York; **Two Clarinets, Close-Up**: © Comstock/CORBIS; **Halley's Diving Bell, 1701**: The Granger Collection, New York; **Piano by Christofori, 1726**: The Granger Collection, New York; **Seed Drill, 18th Century**: The Granger Collection, New York; **Robert Fulton's *Clermont*:** The Granger Collection, New York; **Bridge**: The Granger Collection, New York; **Close-up of an Upright Tuning Fork**: © Stockbyte/CORBIS; **Steam Engine**: The Granger Collection, New York; **Sextant Octant**: Fanny Schertzer, Wikimedia Commons; **Loom: Fly Shuttle, 1733**: The Granger Collection, New York; **Franklin Stove**: The Granger Collection, New York; **Musschen-broek: Leyden Jar**: The Granger Collection, New York; **Franklin: Lightning Rod:** The Granger Collection, New York; **Daniel Boone (1734–1820)**: The Granger Collection, New York; **Check**: Vladimir Ivanov/123rf; **John Harrison with His Marine Chronometer, c. 1735**: Science Museum Pictorial; **The Spinning Jenny**: The Granger Collection, New York; **Puzzle Pieces on Top of Each Other**: Andres Rodriguez/123rf; **Glass with Sparkling Water**: Lein De Leon/123rf; **Arkwright's Spinning Frame**: The Granger Collection; **Cugnot's Steam Carriage**: The Granger Collection, New York; **Ornate Venetian Face Mask**: Nicola Gavin/123rf; **Georges–Louis Lesage's Original Telegraph**: Public Domain; **Submarine: Turtle, 1776**: The Granger Collection, New York; **Thomas Crapper (1837–1902)**: The Granger Collection, New York; **Orukter Amphibolos, 1804**: The Granger Collection, New York; **Circular Saw Blade**: Dana Bartekoske/123rf; **View from the Terrace of Monsieur Franklin a Passy**: Bridgeman–Giraudon/Art Resource, NY; **Bramah Lock**: Wikimedia Commons; **Benjamin Franklin**: The Granger Collection, New York; **Pair of annular wick Argand lamps, c. 1800:** Science Museum; **Scattershot**: Wikimedia Commons; **"A Peep at the Gas Lights in Pall Mall," London, 1807**: Science Museum Pictorial; **Textile**

Manufacture, 1892: The Granger Collection, New York; **Oliver Evans's Mill**: The Granger Collection, New York; **Meikle's Threshing Machine, c. 1788**: Science Museum; **Claude Berthollet**: The Granger Collection, New York; **Constitution**: The Granger Collection, New York; **Samuel Slater (1768–1835)**: The Granger Collection, New York; **Black Nails**: Andrzej Tokarski/123rf; **"A French Dentist Showing His Artificial Teeth and False Palates," 1798**: Science Museum Pictorial; **18th Century Model of a Guillotine**: Bridgeman –Giraudon/Art Resource, NY; **Semaphore**: Wikimedia Commons; **Spools of Color Thread**: Konrad Lewandowski/123rf; **Whitney Cotton Gin**: The Granger Collection, New York; **Ball Bearing**: Jon Kroninger/123rf; **Nuts and Bolts**: Kirsty Pargeter/123rf; **Wheel-lock Rifles**: Erich Lessing/Art Resource, NY; **First Vaccination, 1796**: The Granger Collection, New York; **Iron Plow**: Wikimedia Commons; **Parachute**: Lisa James/123rf; **Lithography, c. 1840**: The Granger Collection, New York; **Voltaic Pile, 1800**: The Granger Collection, New York; **Jacquard Loom**: The Granger Collection, New York; **Trevithick's Locomotive**: The Granger Collection, New York; **Robert Fulton's *Clermont***: The Granger Collection, New York; **Tin Cans, 2002**: Science Museum Pictorial; **Autographs of All Works by Beethoven**: Erich Lessing/Art Resource, NY; **Kaleidoscope**: Jorge Felix Costa/123rf; **Check Your Heart Beat**: Aman Khan/123rf; **Vintage Petrol–Powered Lamp**: Sasha Radosavljevic/123rf; **William Sturgeon's First Electromagnet, 1825**: Science Museum; **World's First Photograph**: The Granger Collection, New York; **Charles Macintosh, Scottish Industrial Chemist and Inventor, c. 1820**: Science Museum; **Closeup of Bible Written in Braille**: Karin Lau/123rf; **Early Matches, c. 1826–1835**: Science Museum; **Hull of SS *Francis Smith* with Smith's Patent Screw Propeller, 1836**: Science Museum; **Grass Mowing Machine, 1830**: The Granger Collection, New York; **Promontory Point, 1869**: The Granger Collection, New York; **The Great Electromagnet, c. 1830–1850**: Science Museum; **McCormick Reaper c. 1875**: The Granger Collection, New York; **Revolver, 19th Century**: The Granger Collection, New York; **Babbage Computer**: The Granger Collection, New York; **Telegraph Key**: The Granger Collection, New York; **Jacob Perkins (1766–1849)**: The Granger Collection, New York; **John Deere Plow**: The Granger Collection, New York; **Charles Goodyear (1800–1860)**: The Granger Collection, New York; **Postage Stamp, 1840**: The Granger Collection, New York; **Baron Justus von Liebig (1803–1873)**: The Granger Collection, New York; **William T. G. Morton**: The Granger Collection, New York; **Trek in Mountains, Portland, England**: Vaide Seskauskiene/123rf; **Ice Cream Freezer, 1872**: The Granger Collection, New York; **Alfred Nobel (1833–1896)**: The Granger Collection, New York; **Sewing Machine, 1846**: The Granger Collection, New York; **Ignaz Semmelweis (1818–1865)**: The Granger Collection, New York; **Hunt: Safety Pin, 1849**: The Granger Collection, New York; **Amelia Bloomer (1818–1894)**: The Granger Collection, New York; **Elisha Graves Otis (1811–1861)**: The Granger Collection, New York; **Gyroscope, 19th Century**: The Granger Collection, New York; **French Aspirin, c. 1950**: The Granger Collection, New York; **Bunsen Burner**: Lance Rider/123rf; **Celluloid Waterproof Collars, Cuffs and Shirt Bosoms, c. 1885**: Science Museum; **Closeup of Rusty Steel Beams**: Cecilia Lim/123rf; **Louis Pasteur in His Laboratory**: Bildarchiv Preussischer Kulturbesitz/Art Resource, NY; **Perkin's Original Mauve Dye, 1856**: Science Museum; **Mason Jar**: The Granger Collection, New York; **Lenoir Gas Engine, 1860**: Science Museum; **Vincent Van Gogh (1853–1890) Skull with Cigarette, 1885**: Art Resource, NY; **Pennsylvania Oil Well**: The Granger

Collection, New York; **Black and White Checkered Floor**: Kim Hall/123rf; **Padlock and Key**: Sean Gladwell/123rf; **Monitor vs. Merrimack**: The Granger Collection, New York; **Gatling Gun, 1867**: The Granger Collection, New York; **Winchester 1872**: Bob Adams/ Wikimedia Commons; **Antique Typewriter**: Andrew Watzenboeck/ 123rf; **George Westinghouse (1846–1914)**: The Granger Collection, New York; **Scooping Butter with a Knife**: © Wim Hanenberg/ Imageshop/CORBIS; **Gumballs**: James Lewis/123rf; **Stock Ticker, 1885**: The Granger Collection, New York; **Closeup of One Strand of Green Metal Barbed Wire**: Marilyn Barbone/123rf; **Thames Embankment, 1867**: Science Museum Library; **Bissell Maid**: Bissell Homecare, Inc.; **Facsimile of Alexander Graham Bell's Original Telephone**: Giraudon/Art Resource, NY; **Thomas Edison (1847–1931)**: The Granger Collection, New York; **Swan's Electric Filament Lamp, 1878–1879**: Science Museum; **Edison's Lightbulb, 1879**: The Granger Collection, New York; **Fannie Merritt Farmer (1857–1915)**: The Granger Collection, New York; **Nikola Tesla (1856–1943)**: The Granger Collection, New York; **Old-Fashioned Cash Register**: James Steidl/123rf; **Maxim with Machine Gun, c. 1880s**: Science Museum Library; **Index Finger**: Michael Pettigrew/123rf; **Burroughs Registering Accountant, c. 1900**: Science Museum; **J. K. Starley, Cycle Engineer, on an Early Rover Cycle, c. 1870s–1879**: Science Museum Pictorial; **Engine**: Andrew Watzenboeck/123rf; **Daimler Motorcycle, 1885**: Science Museum; **First Benz Auto, 1886**: The Granger Collection, New York; **Camera: The Kodak #1, 1889**: The Granger Collection, New York; **Escalators**: Dimitry Romanchuck/123rf; **Red and Blue Zippers**: Marta Menéndez/ 123rf; **Several Old Tractors**: Denis Pepin/123rf; **Edison Kinetoscope, 1894**: NMPFT; **Gugliemo Marconi, Italian Radio Pioneer, c. 1900**: Science Museum; **Compact Cassette**: Andrzej Tokarski/123rf; **Blue Paper Clip**: Mariano Ruiz/123rf; **Graf Zeppelin in Flight**: The Granger Collection, New York; **Gillette Razor Ad, 1906**: The Granger Collection, New York; **"Friends," Poster Promoting the British Vacuum Cleaning Company Ltd., 1906**: Science Museum; **Set of Generic Conditioners on a Wall**: Fedor Sidorov/123rf; **Wright Brothers, 1903**: The Granger Collection, New York; **Standing Tea Bag**: Marc Dietrich/123rf; **Car Covered with Snow**: © Perry Mastrovito/CORBIS; **Advertisement: Washer**: The Granger Collection, New York; **Autochrome Lumiere 1900s**: The Granger Collection, New York; **Bakelite Powder Box, 1921–1940**: Science Museum; **Gyrocopmpass**: Wikimedia Commons; **Early Stainless Steel Tea Knife, c. 1915**: Science Museum; **Crossword Puzzle**: Adrian Hughes/ 123rf; **Coolidge X-ray Tube, 1913–1923**: Science Museum; **Physics: Apparatus, 1879**: The Granger Collection, New York; **Automobile Manufacturing**: The Granger Collection, New York; **Paul Langevin (1872–1946)**: The Granger Collection, New York; **Machine Carbine with Bullet:** Gennady Kravetsky/123rf; **Bandage**: Byron Moore/123rf; **Toaster Ad, 1931**: The Granger Collection, New York; **Television, 1950s**: The Granger Collection, New York; **Frozen Food Ad**, 1957: The Granger Collection, New York; **Aerosol Spray Can with Hand**: Ablestock Premium/123rf; **Movie Theater, 1927**: The Granger Collection, New York; **Bread and Oats**: Dana Bartekoske/123rf; **Schick Electric Razor with Spare Cutter, c. 1934**; Science Museum; **Robert Hutchings Goddard (1882–1945)**: The Granger Collection, New York; **Ernest Orlando Lawrence (1901–1958)**: The Granger Collection, New York; **Technicolor Three–Color 35mm Camera, American, 1932–1955**: NMPFT; **Edwin Armstrong:** National Oceanic and

Atmospheric Administration (NOAA); **Sir Robert A. Watson–Watt (1892–1973)**: The Granger Collection, New York; **Juan de la Cierva (1896–1936)**: The Granger Collection, New York; **Whittle W1 Jet Propulsion Engine, 1941–1944**: Science Museum; **Triumph Velure Fully-Fashioned Stockings, 1950–1965**: Science Museum; **Asian Woman Exercising**: 123rf; **Biro pen**: Daniel Schwen/ Wikimedia; **Fiberglass Light on Black**: Gennady Kudelya/123rf; **Photocopy Machines, 1950**: The Granger Collection, New York; **Ampules of Penicillin, 1943**: NMPFT *Daily Herald* Archive; **Nuclear Reactor, 1942**: The Granger Collection, New York; **Jacques Cousteau (1910–1997)**: The Granger Collection, New York; **Kidney Machine Used for Home Dialysis, c. 1966**: Science Museum; **Harvard Mark I Computer**: Topory/Brudnopis; **Castle Romeo**: Department of Energy; **Microwave Oven**: Sascha Burkard/123rf; **Vacuum Tube Closeup**: Jozsef Szasz-Fabian/ 123rf; **Stack of Diapers**: Eric Isselée/123rf; **Candida Albicans Yeast**: © Visuals Unlimited/CORBIS; **Old Vinyl Record (LP)**: Andrzej Tokarski/123rf; **Polaroid Land Camera, Model 95, 1948–1953**: NMPFT; **Pair of White Training Shoes**: Adrian Hughes/123rf; **Barcode**: Melissa King/123rf; **Cesium Atomic Clock, 1955**: Science Museum; **Credit Card—Financial Background**: Olga Aleksandrovna Lisitskaya/123rf; **Collection of Early Contraceptive Pills, 1960–1980**: Science Museum; **Edward Teller**: Lawrence Livermore National Laboratory/Wikimedia; **Silver Retro Radio**: James Steidl/123rf; **Solar Panels**: Robert Fullerton/123rf; **Pancreatic Cancer Cells**: Dr. Lance Liotta Laboratory/National Cancer Institute; **Antibiotic**: Marek Kosmal/123rf; **Jonas Salk (1914–1995)**: The Granger Collection, New York; **The SR–N1, the World's First Hovercraft, Making Maiden Voyage, 11 June 1959**: NMPFT *Daily Herald* Archive; **Ampex Videotape Recorder, Type VR1000A, Serial Number 329, c. 1950s**: Science Museum; **Sputnik**: NASA/Wikimedia Commons; **Unimate 2000B Industrial Robot, c. 1979**: Science Museum; **Competitions on Snowmobile**: Andrey Stratilatov/ 123rf; **Riber Optic Cables**: Ablestock Premium/123rf; **Wavy Rainbow Laser Light**: Masterpiece Zakariatengah/ 123rf; **Close–up of a Tube of Super Glue**: © Stockdisc/CORBIS; **Widney Motor Vehicle Seat Belt, 1955–1965**: Science Museum; **Integrated Circuit**: Ew Chee Guan/123rf; **First Pacemaker**: © Bettmann/CORBIS; **The First Transatlantic TV Picture from Space, 1962**: NMPFT *Daily Herald* Archive; **Yuri Gagarin, Russian Cosmonaut, 1961**: Science Museum; Pictorial; **Light Emitting Diodes**: Piccolo Namek; **Automated Teller Machine**: Aaron Kohr/123rf; **Pocket Calculator**: Alessandro Bolis/123rf; **Hamilton Pulsar Digital Wristwatch, 1970**: Science Museum; **Computer Aided Magnetic Tomography of Head, Lateral View**: Danilo Ascione/123rf; **Apple II Micro Computer, 1977**: Science Museum; **NEC Biodegradable Phone, c. 2006, and Older Phone**: Science Museum; **Colorful Sticky Notes**: Dana Bartekoske/123rf; **Early Compact Disc Player, 1983**: Science Museum; **Prozac Foil Pack**: Michiel 1972/ Wikimedia Commons; **Tim Berners-Lee, Pioneer of the World Wide Web, c. 1990s**: CERN; **Amazon.com Chairman Jeff Bezos Speaks at Shareholders Meeting**: © Reuters/CORBIS; **Dolly the Cloned Sheep**: © Karen Kasmauski/CORBIS; **iPod shuffle**: H. Rotgers/123rf; **Wikipedia logo**: Wikimedia Foundation.

Index

Also available in **THE ILLUSTRATED TIMELINE** series:

The Illustrated Timeline of Art History

The Illustrated Timeline of Science

The Illustrated Timeline of The Universe